DATE DUE

Refining of Synthetic Crudes

Martin L. Gorbaty, EDITOR

Exxon Research and Engineering Company

Brian M. Harney, EDITOR

U.S. Department of Energy

Based on a symposium

sponsored by the Division of

Petroleum Chemistry at the

174th Meeting of the

American Chemical Society,

Chicago, Illinois,

August 29–September 1, 1977.

ADVANCES IN CHEMISTRY SERIES **179**

AMERICAN CHEMICAL SOCIETY

WASHINGTON, D. C.   1979

Library of Congress CIP Data

Refining of synthetic crudes.
   (Advances in chemistry series; 179 ISSN 0065–2393)

   Includes bibliographies and index.

   1. Petroleum, Synthetic—Congresses. 2. Petroleum—Refining—Congresses.
   I. Harney, Brian M., 1944– . II. Gorbaty, Martin L., 1942– . III. American Chemical Society. Division of Petroleum Chemistry. IV. Series.

QD1.A355 no. 179   [TP355]   540'.8s   [662'.662]
                                          79–21098
   ISBN 0–8412–0456–X   ADCSAJ 179 1–218 1979

FOREWORD

ADVANCES IN CHEMISTRY SERIES was founded in 1949 by the American Chemical Society as an outlet for symposia and collections of data in special areas of topical interest that could not be accommodated in the Society's journals. It provides a medium for symposia that would otherwise be fragmented, their papers distributed among several journals or not published at all. Papers are reviewed critically according to ACS editorial standards and receive the careful attention and processing characteristic of ACS publications. Volumes in the ADVANCES IN CHEMISTRY SERIES maintain the integrity of the symposia on which they are based; however, verbatim reproductions of previously published papers are not accepted. Papers may include reports of research as well as reviews since symposia may embrace both types of presentation.

CONTENTS

PREFACE

It now is recognized that increasing amounts of liquid fuels will have to be derived from sources other than petroleum, such as shale, tar sand bitumen, and coal. Generally, the quality of a fuel may be correlated with its atomic hydrogen-to-carbon (H/C) ratio. For example, methane has an atomic H/C ratio of 4 while crude petroleum and coal have ratios of about 1.7 and 0.8, respectively. Thus, the upgrading of coal to a liquid may be viewed as a hydrogen addition procedure. Also, concentrations of atoms other than carbon and hydrogen in coal-, shale-, and bitumen-derived liquids such as nitrogen, sulfur, oxygen, and trace inorganics, have to be lowered before they can be processed using current petroleum-refining technology.

The chapters in this book encapsulate a review of the state of the art in refining and upgrading of synthetic crudes, i.e., liquids derived from coal, oil shale, and tar sand bitumen. The first several chapters set the challenges in perspective by helping to define the molecular composition of liquids derived from coal. Following chapters discuss refining of shale oil and tar sand bitumen. The last chapters describe a number of different approaches directed toward the upgrading and refining of liquids derived from coal.

The chapters in this volume were presented originally at the national meeting of the American Chemical Society in August of 1977; many of them have been revised and updated. We wish to thank the contributing authors for their efforts.

It is our belief that this volume can serve as a source of information and a base from which future science and technology thrusts can emerge.

MARTIN L. GORBATY
Exxon Research and
    Engineering Company
P.O. Box 45
Linden, New Jersey 07036
August 6, 1979

BRIAN M. HARNEY
Department of Energy
Office of Shale Research Applications
12th and Pennsylvania Ave., N.W.
Washington, D.C. 20461

Advances in Chemistry Series

M. Joan Comstock, *Series Editor*

Characterization Data for Syncrudes and Their Implication for Refining

J. E. DOOLEY, W. C. LANNING, and C. J. THOMPSON

Bartlesville Energy Research Center, U.S. Department of Energy, P.O. Box 1398, Bartlesville, OK 74003

The character and hydrocarbon-type composition of several syncrudes have been investigated by adaptation of methods developed for heavier fractions of petroleum crude oils. The methods are reviewed briefly, and results are summarized for five coal liquids and a hydrotreated shale oil. Refining requirements for removal of heteroatoms, especially nitrogen, and conversion of polynuclear aromatics are discussed in relation to the composition of the syncrudes and the character of refined products to be expected. A preliminary report is given on the preparation of liquid samples from coals of widely different rank to permit more systematic correlation of hydrocarbon character with coal source in relation to refining.

Detailed characterization data are needed to realize the maximum potential of coal liquefaction products, shale oil, and tar sand liquids. Coal conversion processes (1, 2) now in development stages produce a wide variety of products that may or may not be directly amenable to current refining technology. The high nitrogen and aromatic contents of coal liquids have a profound effect on the selection of upgrading processes necessary to convert these materials to finished products. Nitrogen compounds, for example, are strong poisons for catalysts in both catalytic cracking and hydrocracking of heavy distillate feedstocks to produce refined liquid fuels. Aromatics, especially of the fused-ring type, lead to excessive yields of coke in either catalytic cracking or delayed-coking processes. Shale oil presents somewhat similar problems with high nitrogen but less aromatic contents.

Characterizations of synthetic liquids by different workers have been difficult to compare because of varied cut points, different methods of analysis, and incomplete reported data. More systematic data will be needed to characterize these liquids as alternative crude oils and to relate the source and liquefaction process to future requirements for production of refined liquid fuels.

The Bartlesville Energy Research Center (BERC) is working on these problems in order to supply needed characterization data, using systematic procedures (3) developed for characterizing distillable petroleum fractions. The procedures were developed through years of petroleum experience in cooperation with the American Petroleum Institute (API) and the Laramie Energy Research Center (LERC). Other reported methods of analysis (4, 5) are effective in providing similarly separated fractions, but the amount of detail obtained from these fractions by mass spectrometry and other techniques has been limited. In cases where mass spectral group-type methods (6, 7) based on high-ionizing voltage fragmentation patterns have been used for characterization, adequate qualitative and quantitative measurements needed in some applications are not possible. Mass spectrometry (MS) has been used (8) on fractions derived from minimal separation of components to provide excellent detail, but it lacks the capability to distinguish between certain hydrocarbon classes that overlap in mass spectral Z series distributions. The techniques developed and used at BERC take advantage of gradient elution chromatographic separation of distillates into saturate and aromatic concentrates. A given aromatic concentrate, e.g., the diaromatics, is further separated by molecular size by gel permeation chromatography (GPC). The GPC fractions are in turn analyzed by low-resolution, low-voltage MS and correlations established previously between the MS and GPC data. The hydrocarbon types in the given aromatic concentrate are recombined to give the composition in considerable detail. The BERC techniques have been applied with some slight modifications to a number of coal liquids (9, 10, 11, 12) available through development projects now in progress. Data so obtained are summarized and compared in this report according to aromatic and naphthene content in order to suggest to the processor the kinds of hydrocarbons he must cope with for upgrading. Preliminary data developed in a similar manner for a hydrotreated shale oil are included also.

A survey of the literature suggests that the types and complexity of hydrocarbons that can be obtained from a given coal can vary with the rank of coal. The character of the coal liquid can be affected also by liquefaction conditions, making useful comparison of liquids made in different processes, frequently at severe conditions, as well as from different coals, very difficult. In the absence of suitably comparable samples,

BERC has undertaken a systematic study which includes the preparation at mild reaction conditions of liquids from coals which vary widely in rank and quality. The liquids so obtained will be characterized and analyzed for hydrocarbon type by BERC procedures, and the data will be used to supplement the characterization studies in progress on other syncrudes. A preliminary report will be made on the progress of this work to date.

Experimental

The separation and characterization scheme used to generate appropriate compositional data has been discussed elsewhere (3, 9, 10, 11, 12) and only a brief review will be given. First, the full-boiling-range product is cut into distillates having boiling ranges approximating < 200°C, 200–370°C, 370–540°C, and 540°C + residuum. More recent modifications (to meet the expected needs of refining studies) provide distillates of more precise determination with boiling ranges of ∼ < 200°C, 200–325°C, 325–425°C, 425–540°C, and 540°C + residuum. Those distillates boiling below 200°C are separated by use of high-pressure liquid chromatography (HPLC) and the fractions are analyzed by MS. The distillates boiling above 200°C are separated by liquid–solid chromatography into saturate and aromatic concentrates, and the aromatic concentrates are further separated by GPC. Acid and base removal is usually accomplished by chemical extraction of the polyaromatic–polar concentrates obtained from the liquid–solid chromatographic separation. Fractions produced in all of these separations are analyzed by MS (and GPC correlations), UV fluorescence spectrometry, and nuclear magnetic resonance (NMR) spectrometry to determine the qualitative and quantitative distributions of compound types found in the fractions. Data derived from these techniques can be so detailed as to provide quantitative carbon number distributions for identified compound types, or the detailed data can be recombined in various forms to provide some of the most accurate and reliable summary compositional data available. Summary-type data will be discussed in this report.

For the batch preparation of coal liquids, BERC is using a 2-L magnetically stirred batch autoclave supplied by Autoclave Engineers. The experimental procedures being used are adaptations of those used at the Pittsburgh Energy Research Center (PERC) and by other research groups, and they are not intended to represent development of a practical process. Rather, a liquid which represents most of the hydrocarbons contained in the source coal is sought, with a minimum of cracking of the original hydrocarbons. Hydrogenation conditions generally have been at 2,500–2,700 psig, with the temperature being increased slowly from approximately 300°–400°C over a period of several hours. Hydrogen was added periodically to maintain total pressure. For two coals processed to date, hydrogen consumption started at or below 325°C. A typical charge to the autoclave consisted of 375 g of coal slurried with 500 g of solvent (Tetralin or filtered liquid product recycled in successive runs), 40–60 g of crushed Cyanamid HDS-3A NiMo catalyst, and 1–2 g of sulfur (to help sulfide the catalyst). The coal–solvent slurry, catalyst, and sulfur were introduced through a charging port in the autoclave cover. Product was removed through the same port by means of a dip-tube and vacuum system. Benzene was charged to the autoclave at the end of a series of runs to remove residual product and solids. When sufficient inventory of coal-derived liquid was accumulated, a final upgrading run was made with presulfided catalyst to yield a liquid product which contained less than about 0.4 wt % nitrogen.

Tetralin was used as a hydrogen-donor solvent in the initial stages of liquefaction. For the first coal processed (Illinois No. 6 bituminous), the Tetralin–naphthalene was removed by distillation after sufficient liquid inventory was accumulated to provide coal liquid as the solvent for further batches. The bottoms from the distillation were solid at ambient temperature and were given an upgrading treatment before use as a recycle solvent.

Results and Discussion

Some of the data to be discussed were reported earlier (13, 14) in somewhat different form and are included with data more recently acquired on other alternate fuels to provide comparative characteristics of all the materials examined.

Coal Liquids and Shale Oil. Some properties of five coal liquids and a hydrotreated shale oil are given in Table I, including sulfur, nitrogen, acid and base contents, and distributions by boiling range. Ring-number

Table I.   Some Physical and Chemical Properties

Liquid	Sulfur % Wt	Nitrogen % Wt	Acids % Wt
COED[b] Utah Coal	0.05	0.48	8.17
COED[c] Western Kentucky Coal	0.08	0.23	3.11
Synthoil[d] West Virginia Coal	0.42	0.79	10.40
H–Coal Fuel Oil[f] Illinois No. 6 Coal	0.21	0.44	14.69
H–Coal Syncrude[g] Illinois No. 6 Coal	0.27	0.63	13.81
Hydrotreated Shale Oil	< 0.01	0.04	—

[a] Approximate boiling ranges.
[b] Approximately 10% of this material was not analyzed. 1.50% Heteroatomic compounds determined in addition to acids and bases extracted.
[c] Approximately 8% of this material was not analyzed. 0.3% Heteroatomic compounds determined in addition to acids and bases extracted.

distributions of the major hydrocarbon classes found in each coal liquid and the hydrotreated shale oil are shown in Tables II, III, IV, and V. While these are only summary data, they represent cumulative totals derived from more detailed data and therefore should be considerably more accurate than those acquired by more gross compositional methods.

Endpoints for the H–Coal syncrude and the two Char Oil Energy Development (COED) syncrudes were determined in earlier work (9, 10, 12) to be in the range of 510°–550°C, but the H–Coal fuel oil endpoint (12) was determined to be lower at approximately 450°–460°C. In Table I, the Synthoil product shows 25.7% residuum boiling above 540°C. These differences in cut points in the liquids make comparisons of data difficult, since the bottoms of the H–Coal materials were not included and the Synthoil residuum was not analyzed and cannot be included in any comparison. However, the nitrogen contents of the coal liquids given in Table I are indicative of the degree of upgrading already done (since the nitrogen concentrations in the coals are high) and can be used as indicators of the degree of upgrading needed to turn these liquids into refined products. For instance, the nitrogen value shown for Synthoil would indicate that this material requires considerably more upgrading than would be necessary for the other liquids shown if part or all of it were to be used for transportation fuels. The larger ring-number systems shown for the Synthoil composite in Tables IV and V tend to substantiate this

for Five Coal Liquids and a Hydrotreated Shale Oil

	Distillate			
Bases % Wt	< 200°C % Wt	200–370°C % Wt	370–540°C % Wt	540°C Resid % Wt
1.30	13.3	45.4	40.3	—
1.05	21.0	54.2	24.2	—
2.75	4.4	42.6	27.3	25.7 [e]
1.29	35.5	63.8		—
2.73	34.6	65.4		—
—	24.3	48.4	24.5	—

[d] Approximately 4% (plus resid) of this material was not analyzed. 2.75% Heteroatomic compounds determined in addition to acids and bases extracted.
[e] Not analyzed.
[f] Only overhead products analyzed. Material above 200°C boiled up to 460°C.
[g] Only overhead products analyzed. Material above 200°C boiled up to 510°C.

Table II.   Ring-Number Distributions[a] for Saturate Concentrates
Derived from Five Coal Liquids and a Hydrotreated Shale Oil

Total Rings	COED UT % Wt	COED W. KY % Wt	Synthoil WV % Wt	H–Coal Fuel Oil IL No. 6 % Wt	H–Coal Syncrude IL No. 6 % Wt	Hydro- treated Shale Oil % Wt
0	8.73	4.26	1.87	1.73	1.42	21.4
1	2.48	4.34	2.62	2.04	1.14	13.3
2	2.15	3.89	1.76	1.90	1.05	8.5
3	2.20	3.16	1.75	1.15	0.62	5.4
4	2.25	2.06	1.15	0.60	0.45	2.7
5	3.96	0.93	0.31	—	—	2.2
6	1.30	0.44	—	—	—	—
< 200°C Sat.	7.05	14.34	1.19	14.20	15.99	22.8
Totals	30.12	33.42	10.65	21.62	20.67	76.3

[a] Determined from distillates in Table I.

observation. This suggests that less carbon–carbon bond cracking occurred in production of the Synthoil liquid, although the coal rank could have some influence. From the distributions shown in Tables III, IV, and V for the coal liquids, aromatics having four or more total rings amount to 24.1, 21.6, 27.0, 8.1, and 18.8% of the total liquid for the Utah COED, Western Kentucky COED, Synthoil, H–Coal fuel oil, and H–Coal syncrude, respectively. This suggests that the Synthoil product contains more of the higher ring systems, but the implication may not be valid since the residuum fractions of the liquids were not analyzed. The shale oil contains only 22.2% total aromatics and these contain only 1–3 rings. Most of the

Table III.   Ring-Number Distributions[a] for Monoaromatic
Concentrates Derived from Five Coal Liquids
and a Hydrotreated Shale Oil

Total Rings	COED UT % Wt	COED W. KY % Wt	Synthoil WV % Wt	H–Coal Fuel Oil IL No. 6 % Wt	H–Coal Syncrude IL No. 6 % Wt	Hydro- treated Shale Oil % Wt
1	1.75	2.04	0.30	1.15	0.96	1.9
2	5.10	8.22	0.95	7.85	4.07	4.4
3	4.79	9.23	4.84	5.02	2.12	3.5
4	3.12	5.67	4.85	0.61	0.35	2.1
5	1.74	2.42	1.57	0.11	0.10	1.0
6	0.56	0.88	0.37	0.07	0.04	0.3
7	0.16	0.32	0.10	—	0.02	0.2
8	0.05	0.11	0.04	—	0.02	trace
< 200°C Monoaromatics	2.79	4.75	1.21	12.18	10.85	4.0
Totals	20.06	33.64	14.23	26.99	18.53	17.4

[a] Determined from distillates in Table I.

Table IV. Ring-Number Distributions[a] for Diaromatic Concentrates Derived from Five Coal Liquids and a Hydrotreated Shale Oil

Total Rings	COED UT % Wt	COED W. KY % Wt	Synthoil WV % Wt	H–Coal Fuel Oil IL No. 6 % Wt	H–Coal Syncrude IL No. 6 % Wt	Hydrotreated Shale Oil % Wt
2	4.13	2.14	3.26	6.49	4.15	1.1
3	4.32	3.83	4.62	5.13	3.40	1.1
4	2.91	3.15	3.19	1.47	1.74	0.6
5	1.81	1.94	1.56	0.43	0.79	0.3
6	0.77	1.04	1.18	0.04	0.31	0.1
7	0.34	0.48	0.99	0.03	0.22	0.1
8	0.13	0.16	0.36	trace	0.04	trace
9	—	0.04	0.16	—	0.04	trace
10	—	—	0.05	—	0.02	trace
11	—	—	0.02	—	—	trace
< 200°C Diaromatics	—	—	0.14	trace	0.06	—
Totals	14.41	12.78	15.53	13.59	10.77	3.3

[a] Determined from distillates in Table I.

coal liquids are highly aromatic (saturates ranged from 11–33% of the liquid analyzed) and contain significant amounts of heteroatomic compound types such as acids (3.1–14.7%) and bases (1.1–2.8%). On the other hand, the hydrotreated shale oil is very high in saturates (76%). Furthermore, the saturate fractions of the coal liquids boiling above 200°C contain approximately 63–80% naphthenes, part of which would be aromatics in the original coal.

Table V. Ring-Number Distributions[a] for Polyaromatic Concentrates Derived from Five Coal Liquids and a Hydrotreated Shale Oil

Total Rings	COED UT % Wt	COED W. KY % Wt	Synthoil WV % Wt	H–Coal Fuel Oil IL No. 6 % Wt	H–Coal Syncrude IL No. 6 % Wt	Hydrotreated Shale Oil % Wt
3	1.94	1.48	1.03	5.81	3.76	0.4
4	3.65	2.11	2.93	3.84	4.15	0.5
5	3.63	1.62	2.21	0.97	3.69	0.3
6	3.58	1.25	2.18	0.32	3.84	0.2
7	1.67	1.24	2.28	0.07	0.81	0.1
8	0.01	0.18	2.04	0.07	1.41	trace
9	—	0.03	1.05	0.03	0.67	trace
10	—	—	0.70	—	0.39	—
11	—	—	0.19	—	0.19	—
Totals	14.48	7.91	14.61	11.11	18.91	1.5

[a] Determined from distillates in Table I.

The ring-number tabulations in Tables II, III, IV, and V are satisfactory for a simplified summary of composition and comparison. The typical Co–Mo or Ni–Mo catalyst commonly used for upgrading will saturate many of the multiple aromatic rings, depending upon severity and activity, but frequently not the last ring of a condensed-ring polyaromatic. Thus, the total number of rings is a measure of the complexity of the hydrocarbon structure. As noted earlier, more detailed data on the distribution of hydrocarbon types in these liquids are available when needed.

Batch Preparation of Coal Liquids at BERC. Two coal liquids had been prepared at the time this report was written, but preliminary analytical data for only one were available and these data are shown in Table VI for the coal liquid prepared from Illinois No. 6 coal. Also shown in the table are typical analytical data for the original coal as supplied by E. L. Obermiller of Conoco Coal Development Co. Mr. Obermiller also provided samples of four other coals used in this program. A sixth sample (lignite) was supplied by J. E. Schiller of the Grand Forks Energy Research Center. While the data in Table VI are only preliminary, sufficient experience has been gained with the batch preparation of small samples of liquids at BERC to suggest that the original goal can be realized to establish correlations of the complexity of the hydrocarbons derived from a given coal with its rank or quality. The three distillates prepared from the Illinois No. 6 coal appear to be quite stable; the less than 200°C distillate is completely clear of any coloration and the 200°–325°C distillate is only slightly yellow in color after about three weeks on the shelf. Asphaltenes will be removed from the 425°C residuum and

Table VI. Preliminary Data for Coal Liquid Prepared from Illinois No. 6 Coal and Typical Properties of Original Coal

Upgraded Coal Liquid		*Typical Analyses of Coal, % Wt*	
Specific Gravity	1.006	Moisture	7
SSU Viscosity @ 100°F	129 sec	Volatile Matter	38.15 [a]
SSU Viscosity @ 130°F	65 sec	Ash	6.75 [a]
Sulfur, % Wt	0.03	Sulfur	1.20 [a]
Nitrogen, % Wt	0.444	Nitrogen	1
Carbon, % Wt	90.1	Carbon	69
Hydrogen, % Wt	9.88	Hydrogen	5
Distillates Prepared:			
< 200°C, % Wt	12.3		
200°–325°C, % Wt	27.3		
325°–425°C, % Wt	20.7		
> 425°C Resid, % Wt	39.7		

[a] Actual analysis.

the asphaltene-free residuum will be vacuum flashed to 540°C to provide a heavier distillate fraction. These distillates are suitable for and will be processed through our separation and characterization scheme to establish hydrocarbon types and quantities.

Applications to Refining

The first essential step in refining these synthetic crude oils is severe catalytic hydrogenation to remove most of the sulfur, nitrogen, and oxygen and to convert at least part of the heavy ends to lower-boiling distillates for further refining. The removal of nitrogen is more difficult than that of sulfur or oxygen and is, therefore, crucial to the early stages of refining. The hydrocarbon character of shale oil already resembles that of many petroleum crude oils, as indicated by its hydrogen/carbon atom ratio of approximately 1.6, except that it has been cracked thermally and typically contains 2.0% nitrogen. The crude shale oil can be subjected to a preliminary delayed coking in order to convert residuum and decrease the required severity of subsequent hydrogenation to remove nitrogen. However, this would not be suitable for highly aromatic coal liquids because the heavy ends would yield little but coke and gas. Another requirement in the initial upgrading by hydrogenation is coping with residual solids in the synthetic crude oil, which thus far has been present typically at concentrations of 0.1% or more of the synthetic crude oil from coal, or about ten times as much as in petroleum residua being charged to hydrodesulfurization processes.

However, the really limiting factor in the refining of coal liquids is attributable to their highly aromatic character as shown in Tables II, III, IV, and V in particular to the presence of substantial concentrations of condensed-ring polyaromatics such as chrysenes, pyrenes, etc. Without going into detail, each condensed aromatic ring must be saturated by hydrogenation before it can be cracked to produce lower-boiling distillate products. Thus, hydrogen consumption increases by large increments as the number of aromatic rings in the condensed-ring hydrocarbon increases, and more severe process conditions are required. Furthermore, the difficulty of such conversion increases sharply for the more highly condensed aromatics of the pyrene type, which are slow to react and tend to cover catalysts and prevent conversion of other heavy components (15). Nitrogen is incorporated in these essentially aromatic ring structures, which explains why it is more difficult to remove than sulfur or oxygen. In compounds with a large number of condensed rings, as in acridine and higher homologs, the nitrogen-containing ring is still less accessible for saturation and cracking for removal of the nitrogen. Aromatic rings are extremely resistant to the cracking, witness hydrodealkylation process which converts alkylbenzenes to benzene at temperatures of 1150°–1300°F (621°–704°C) with no loss of rings. Sulfur and oxygen, being divalent, are excluded from such stable benzene-like rings and are in compounds from which they are more easily removed by hydrogenation. Coal liquids produced by noncatalytic liquefaction, even in the presence of hydrogen, contain essentially the same 1.0–1.4% nitrogen as the original. In fact, the degree to which a synthetic crude oil from coal has been upgraded by catalytic hydrogenation can be fairly well gaged by the nitrogen content—typically about 0.5% as illustrated in Table I for relatively moderate hydrotreatment to 0.05% for severe hydrotreatment. Removal of nitrogen from the heavier ends of synthetic crude oils is required for hydrocracking to lighter distillates and gasoline, although petroleum hydrocracking processes are reported to tolerate as much as 0.2–0.3% nitrogen. The two factors of (a) condensed-ring aromatic structures which (b) contain nitrogen account largely for the reported difficulty of conversion, even by hydrogenation and hydrocracking, of the heavy ends of coal liquids (16).

The only detailed refining study reported (16) for a syncrude from coal, using present-day catalytic processes, was made on a severely hydrotreated liquid from the COED pyrolysis process. This feedstock may not be fully representative of syncrudes which can be produced by direct hydrogenation of the coal to give much greater percentage yields of liquids, but the difficulties encountered in converting the heavy ends boiling above about 750°F (399°C) to lighter distillate products, for reasons which have already been discussed, suggest that such heavy ends may be blended better into heavy fuel oil than to add excessively to the cost of refining.

The lighter distillates and naphtha from oil shale and coal can be refined by processes used for petroleum, except that more severe hydrotreatment will be required to remove nitrogen and the other nonhydrocarbon impurities that poison catalysts and cause product instabiltiy.

Once the synthetic crude oils from coal and oil shale have been upgraded and the heavy ends converted to lighter distillates, further refining by existing processes need not be covered in detail except to note the essential character of the products. The paraffinic syncrude from oil shale yields middle distillates which are excellent jet and diesel fuel stocks. The principal requirements are removal of nitrogen to the extent necessary for good thermal stability of the fuels and adjustment of cut points to meet required pour or freeze points, limited by the presence of waxy straight-chain paraffins. The heavy naphtha from shale oil can be further hydrotreated and catalytically reformed to acceptable octane number, but with considerable loss of volume because of the only moderate content of cyclic hydrocarbons, typically 45–50%. On the other hand, syncrudes from coal will contain 75–95% of naphthenic and aromatic hydrocarbons. Accordingly, the heavy naphtha from coal after hydrotreatment is excellent feedstock for catalytic reforming and yields over 90 vol % of 100 octane number (Research, unleaded) gasoline at mild process conditions. The middle distillates from coal would not meet present-day specifications for either jet or diesel fuels without extensive hydrogenation because of high aromatic content and the virtual absence of straight-chain paraffins.

Conclusions

The methods used at BERC to characterize petroleum and coal liquids provide a means for systematic study of various alternate or synthetic crude oils. The compositional data so produced are useful for producers or refiners of such materials and give them a base for clearer comparison of all liquids produced. Data from the batch preparation of liquids and the subsequent characterization of these liquids will provide a clearer understanding of the effect of coal source on the hydrocarbon composition of the potential liquefaction products. Such information will be necessary for the proper upgrading of coals and coal liquids and will contribute to more efficient processing and end-use of these materials.

Literature Cited

1. "Coal Conversion Activities Picking Up," *Chem. Eng. News* 1975, 53, 24–25.
2. O'Hara, J. B. "Coal Liquefaction," *Hydrocarbon Process.* 1976, 55(11), 221–226.
3. Haines, W. E.; Thompson, C. J. "Separating and Characterizing High-Boiling Petroleum Distillates: The USBM-API Procedure," ERDA-LERC/RI-75/5, ERDA-BERC/RI-75/2, July 1975.
4. Callen, R. B.; Bendoraitis, J. G.; Simpson, C. A.; Voltz, S. E. "Upgrading Coal Liquids to Gas Turbine Fuels. 1. Analytical Characterization of Coal Liquids," *Ind. Eng. Chem., Prod. Res. Dev.* 1976, 15(4), 223–233.
5. Ruberto, R. G.; Jewell, D. M.; Jensen, R. K.; Cronauer, D. C. "Characterization of Synthetic Liquid Fuels," *Am. Chem. Soc., Div. Fuel Chem. Prepr.*, 1974, 19(2), 258–290.
6. Swanziger, J. T.; Kickson, F. E.; Best, H. T. "Liquid Compositional Analysis by Mass Spectrometry," *Anal. Chem.* 1974, 46, 730–734.
7. Robinson, C. J.; Cook, G. L. "Low-Resolution Mass Spectrometric Determination of Aromatic Fractions from Petroleum," *Anal. Chem.* 1969, 41, 1548–1554.
8. Aczel, T.; Foster, J. Q.; Karchmer, J. H. "Characterization of Coal Liquefaction Products by High Resolution-Low-Voltage Mass Spectrometry," *Am. Chem. Soc., Div. Fuel Chem., Prepr.* 1969, 13(1), 8–17.
9. Dooley, J. E.; Sturm, G. P., Jr.; Woodward, P. W.; Vogh, J. W.; Thompson, C. J. "Analyzing Syncrude from Utah Coal," ERDA-BERC/RI-75/7, Aug. 1975.

10. Sturm, G. P., Jr.; Woodward, P. W.; Vogh, J. W.; Holmes, S. A.; Dooley, J. E. "Analyzing Syncrude from Western Kentucky Coal," ERDA-BERC/RI-75/12, Nov. 1975.
11. Woodward, P. W.; Sturm, G. P., Jr.; Vogh, J. W.; Holmes, S. A.; Dooley, J. E. "Compositional Analyses of Synthoil from West Virginia Coal," ERDA-BERC/RI-76/2, Jan. 1976.
12. Holmes, S. A.; Woodward, P. W.; Sturm, G. P., Jr.; Vogh, J. W.; Dooley, J. E. "Characterization of Coal Liquids Derived from the H-Coal Process," ERDA-BERC/RI-76/10, Nov. 1976.
13. Dooley, J. E.; Thompson, C. J. "The Analysis of Liquids from Coal Conversion Processes," *Am. Chem. Soc., Div. Fuel Chem., Prepr.,* **1976**, *21*(5), 243–251.
14. Lanning, W. C.; Thompson, C. J. "Alternative Liquid Fuels and the ERDA Program," ASME Pet. Div. Prepr. No. 76-PET-15, 1976.
15. Hogan, R. J.; Mills, K. L.; Lanning, W. C. "Hydrogenation of Some Condensed-Ring Aromatic Hydrocarbons," *J. Chem. Eng. Data* **1962**, *7*(1), 148–149.
16. Jones, J. F. "Evaluation of COED Syncrude," Office of Coal Research R&D Rept. No. 73, Interim No. 3, Appendix 1, March 1975.

RECEIVED June 28, 1978.

Mass Spectrometric Analysis of Heterocompounds in Coal Extracts and Coal-Liquefaction Products

THOMAS ACZEL and H. E. LUMPKIN

Exxon Research and Engineering Company, Baytown Research and Development Division, Analytical Research Laboratory, Baytown, TX 77520

A large number of heteroaromatic components (carbon-number homologs) have been determined in coal liquids. Analytical techniques used were principally high-resolution and ultra-high-resolution low-voltage mass spectrometry, IR, NMR, UV, and separation techniques. Emphasis is placed on the unique capability of ultra-high-resolution mass spectrometry to determine the formulas of essentially all heteroaromatics present in the materials examined. A summary list of more than 1200 formulas determined in the authors' laboratory includes one-to-eight ring condensed furans, difurans, trifurans, thiophenes, dithiophenes, thiophenofurans, hydroxyaromatics, dihydroxyaromatics, hydroxyfurans, pyrroles, pyridines, hydroxy nitrogen compounds, and miscellaneous heterocompounds with up to five heteroatoms per molecule.

Coal-liquefaction products generally contain high concentrations of aromatic heterocompounds. Most of these components adversely affect refining processes that must be designed to upgrade product quality, and if not eliminated, the quality of the final products. Therefore, it is of the utmost interest to develop analytical methodology that can yield very detailed information on both the amount and the structures of these aromatic heterocompounds. Ideally, such methods need to be equally applicable to coal extracts and to any products or by-products of coal liquefaction and gasification and to other conversion

0-8412-0456-X/79/33-179-013$05.00/1
© 1979 American Chemical Society

processes. In practice, these very severe requirements cannot be met by one analytical tool or method, and a combination of several techniques is needed even to approach these goals.

High-resolution mass spectrometry (MS) is one of the analytical techniques that can furnish significant information on the heterocompounds in coal liquids. This technique has been widely applied to the analysis of coal-liquefaction products (*1, 2, 3, 4*) and is essentially indispensable for the detailed analysis of heterocompounds (*3, 5, 6*).

This chapter will discuss the capabilities and the limitations of high-resolution MS in view of the above-discussed requirements, illustrating the conclusions with data obtained mainly in our laboratories. Other techniques will be surveyed also, but only in a very cursory manner, and only to indicate possible alternatives to compensate for the deficiencies of high-resolution MS.

Discussion

Scope of High-Resolution MS. High-resolution MS can be used for the analysis of all types of complex mixtures of heterocompounds that can be volatilized and/or ionized in the instrument. Materials with boiling points up to approximately 650°C can be volatilized from conventional inlet systems furnished with reservoirs in which the volatile components reach equilibrium before the spectrum is scanned. Materials with somewhat higher boiling points can be vaporized using a direct insertion probe into the ionization source, both in an electron-impact and field-ionization mode or using the field desorption technique (*7*); but in these cases there may be a rapid change in the composition of wide-boiling mixtures while the sample is being vaporized. Although this change in composition can be minimized with the use of rapid, replicate scans, multiple-channel detector systems, or photoplate detection, spectra obtained from a conventional inlet system are generally more reproducible and can be run at a higher resolving power.

As both reproducibility and high resolving power are essential for the detailed characterization of complex mixtures of heterocompounds, in most of our work we prefer conventional inlet systems. There are provisions to measure accurately the nonvolatile residue of the material charged (*8*), and there is the possibility to charge sufficient material, up to 200–300 mg, so that one can obtain good quality spectra even if the volatiles amount only to 1% or less of the total material. A typical example of this approach is the determination of parts per million aromatics and heteroaromatics in coal powders that we carried out in our work on the characterization of Synthoil feed and products (*3*).

The availability of high resolution is the most critical parameter in the mass spectrometric determination of heterocompounds. Although resolving-power requirements may be mitigated by extensive separations and by very precise mass measurements in some cases, the complexity of coal-derived mixtures makes the availability of high resolving power a prime consideration. This complexity is well illustrated by some of our findings in the characterization of Synthoil feed and products (3). Highly separated four-plus ring aromatic fractions that underwent successive cyclohexane extraction and clay-gel and alumina column separations, still contain aromatic thiophenes, dithiophenes, furans, difurans, and thiophenofurans in addition to hydrocarbons, and as many as eight peaks of different composition at one nominal mass.

A theoretical multiplet listing most of the molecular ions that could occur in the high-resolution spectrum of a typical coal liquid is shown in Table I.

Resolving-power requirements at m/e 300 are approximately 10,000–17,000 for hydrocarbon/oxygen compound pairs, 37,000 for nitrogen/hydrocarbon isotope pairs, 88,000 for hydrocarbon/sulfur compound pairs, etc. These requirements increase linearly with increasing molecular weight and are twice as large at m/e 600 than at m/e 300.

When one deals with complex mixtures, it is often advantageous to simplify their spectra to parent peaks either by the use of low-voltage technology (2, 9, 10) or field ionization (11). This simplification results in additional difficulties because of the inherent loss of sensitivity. In some cases sensitivity loss can be compensated for by charging large amounts of sample, up to 10–20 mg.

Table I.   Resolving Power Required to Separate Members of Possible Multiplets at m/e 300 and 301

Formula	Mass	Δ Mass	RP Required
$C_{22}H_{36}$	300.2817	—	—
$C_{21}H_{32}O$	300.2453	0.0364	8,000
$C_{20}H_{28}O_2$	300.2089	0.0364	8,000
$C_{20}H_{28}S$	300.1912	0.0177	17,000
$C_{23}H_{24}$	300.1878	0.0034	88,000
$C_{19}H_{24}SO$	300.1548	0.0330	9,000
$C_{22}H_{20}O$	300.1514	0.0034	88,000
$C_{24}H_{12}$	300.0939	0.0575	5,000
$^{13}CC_{22}H_{24}$	301.1912	—	—
$C_{22}H_{23}N$	301.1830	0.0082	37,000
$^{13}CC_{21}H_{20}O$	301.1548	0.0282	11,000
$C_{21}H_{19}NO$	301.1466	0.0082	37,000

There are very few mass spectrometers that are able to satisfy these requirements. In our work we use either an AEI MS9 high-resolution mass spectrometer, resolving power up to 10–20K in a dynamic scan mode, or more recently an AEI MS50 ultra-high-resolution mass spectrometer, resolving power up to 40–80K at the same conditions. The capability of the MS50 for the resolution of heterocompounds in complex coal liquids is illustrated by the multiplets resolved in the spectra of unseparated heavy distillates from a coal liquid and of a light asphaltene fraction from another coal liquid reported in Tables II and III. MS data on these and the other tables in this chapter were obtained in the low-voltage operation mode.

Tables II and III are presented with slightly different formats to illustrate the scope of the data. Spectra were run at about 40,000 resolving power. Approximately 500 individual carbon-number homologs were identified in the spectrum of the heavy liquid, 1000 in that of the asphaltene (12). Total ionization in both cases was about 1,000,000 so that the homologs with intensity values ca. 100 represent approximately 100 ppm concentration. As indicated in the tables, the 40,000 resolving power was insufficient to separate hydrocarbon/sulfur doublets. The identifications given are based on mass measurement and chemical considerations. This is possible only in some cases. Spectra run at 80,000 resolving power showing separated sulfur/hydrocarbon doublets are reported elsewhere (13, 14).

Qualitative Analysis. High-resolution MS generally can separate and identify the formulas of essentially all heteroaromatic components containing one or more oxygen, sulfur, and nitrogen atoms per molecule, although some difficulties remain in the routine identification of sulfur compounds. The approach used to compensate for the latter difficulty is discussed in the section on quantitative analysis.

Table II.   Multiplets in Ultra-High-Resolution MS of a Heavy Coal Liquid

Intensity	Measured Mass	Error	Formula	Required RP
1,689	284.1558	−0.0007	$C_{22}H_{20}$ [a]	84,000
		−0.0041	$C_{19}H_{24}S$	17,000
241	284.1416	+0.0022	$^{13}CC_{20}H_{17}N$	15,000
1,125	284.1189	−0.0012	$C_{21}H_{16}O$	11,000
292	284.0840	+0.0003	$C_{20}H_{12}O_2$	
476	285.1580	−0.0019	$^{13}CC_{21}H_{20}$	35,000
767	285.1502	−0.0015	$C_{21}H_{19}N$	12,000
197	285.1278	−0.0001	$C_{21}H_{17}O$	23,000
268	285.1144	−0.0009	$C_{20}H_{15}NO$	

[a] Hydrocarbon formula accepted on the basis of mass measurement.

**Table III. Decaplet in Ultra-High-Resolution MS of a
Light Asphaltene Fraction from a Coal Liquid**

Intensity	Theor. Mass	Formula	Possible Structure(s)
188	310.1993	$C_{21}H_{26}O_2$	C_8-dihydroxyfluorene
1,246	310.1721	$C_{24}H_{22}$	C_5-cholanthrene
500	310.1569	$C_{20}H_{22}O_3$	C_6-trihydroxyphenanthrene
3,230	310.1358	$C_{23}H_{18}O$	C_3-hydroxybenzopyrene
273	310.1205	$C_{19}H_{18}O_4$	C_4-trihydroxyfluorenofuran
1,760	310.1028	$C_{19}H_{18}SO_2$	C_5-hydroxydibenzothiopheno-furan
198	310.0816	$C_{22}H_{14}S^a$	C_1-thiophenocholanthrene
	310.0782	$C_{25}H_{10}$	—
156	310.0664	$C_{18}H_{14}SO_3$	C_2-trihydroxypyrenothiol
522	310.0487	$C_{18}H_{14}S_2O$	C_2-hydroxydithiophenoace-naphthene
121	310.0123	$C_{17}H_{10}S_2O_2$	C_1-hydroxyfuranothiophenodi-benzothiophene

[a] Sulfur compound preferred to hydrocarbon because of precise mass measurement (0.000 error for $C_{22}H_{14}S$, 0.0034 for $C_{25}H_{10}$) and because no reasonable aromatic structure exists for a $C_{25}H_{10}$ hydrocarbon.

Table IV summarizes more than 300 homologous series (compound types) that we have identified in the large number of coal liquids examined since the middle sixties (2, 3, 15). Each of the homologous series listed contains up to 20–25 members. Most of these identifications were carried out using computer programs developed in our laboratories (16).

High-resolution, low-voltage MS determines only formulas. Assignment of the structures listed in Table IV, as well as of the structures given for approximately 158 heteroatomic compound types in our work on the characterization of Synthoil feeds and products (3) is based on additional considerations. These include mass spectrometric evidence, such as determining the molecular weight of the first homolog in a homologous series or compound type from the observation of a large number of similar samples, and data from separations and other spectrometric techniques.

In many cases we rely on the clay-gel chromatographic technique (17) (modifications related to applications with coal-derived liquids are given in Ref. 18) that separates aromatic furans and thiophenes from polar compounds such as phenols and thiols. Generally, this simple separation enables us to assign reasonable structures to the oxygen compounds. In our experience these are prevalently phenolic, although significant amounts of furans are also present.

Basic nitrogen compounds such as condensed aromatic pyridines can be separated from neutral or slightly acidic types such as condensed

**Table IV.   Summary of Heteroaromatic Compound Types
Identified in Coal Liquids**

Furans, Difurans, etc.	$C_nH_{2n-2}O$ through $C_nH_{2n-46}O$
	$C_nH_{2n-6}O_2$ through $C_nH_{2n-38}O_2$
	$C_nH_{2n-8}O_3$ through $C_nH_{2n-28}O_3$
Thiophenes,	$C_nH_{2n-6}S$ through $C_nH_{2n-36}S$
Dithiophenes, etc.	$C_nH_{2n-6}S_2$ through $C_nH_{2n-32}S_2$
Thiophenofurans	$C_nH_{2n-10}SO$ through $C_nH_{2n-34}SO$
	$C_nH_{2n-8}SO_2$ through $C_nH_{2n-44}SO_2$
	$C_nH_{2n-22}SO_3$ through $C_nH_{2n-38}SO_3$
Hydroxy compounds,	$C_nH_{2n-6}O$ through $C_nH_{2n-42}O$
Dihydroxy compounds,	$C_nH_{2n-6}O_2$ through $C_nH_{2n-40}O_2$
Hydroxyfurans, etc.	$C_nH_{2n-8}O_3$ through $C_nH_{2n-32}O_3$
Nitrogen compounds (one through	$C_nH_{2n-3}N$ through $C_nH_{2n-47}N$
eight cond. ring pyrroles, pyridines)	
Hydroxy-Nitrogen compounds	$C_nH_{2n-5}NO$ through $C_nH_{2n-41}NO$
	$C_nH_{2n-9}NO_2$ through $C_nH_{2n-37}NO_2$
Miscellaneous Sulfur, Oxygen, and	$C_nH_{2n-x}S_3$, $C_nH_{2n-x}S_2O$,
Nitrogen compounds	$C_nH_{2n-x}NO_2S$, $C_nH_{2n-x}O_4$,
	$C_nH_{2n-x}SO_4$, $C_nH_{2n-x}SN$,
	$C_nH_{2n-x}NO_3$, $C_nH_{2n-x}NO_4$, etc.

aromatic pyrroles with a method recently developed by Sternberg, et al. (*19*) for asphaltene fractions, but that probably can be applied also to polar aromatic materials (*20*).

Separations also enrich the fractions to be analyzed in a desired class of heterocompounds and thus partially overcome the sensitivity limitations inherent in low-voltage MS.

More extensive separations (*21, 22*) provide further detail in some cases, but in our work we prefer to rely mainly on high-resolution MS and keep separations to a minimum. In fact, we will forego even the clay-gel separation in most routine analyses and restrict its use to calibrations and in-depth investigations.

In addition to separations, we use spectrometric techniques such as IR, GC/MS, and GC/UV to gain further structural information. IR measurements provide both qualitative and quantitative information on phenols and pyrroles, and in addition to the types listed in Table IV, on carboxylic acids and aromatic amines. However, we have observed the latter two types only in coal-feed extracts (*3*). Individual heteroaromatics can be detected by GC/MS and GC/UV. Several dibenzofurans, phenols, indoles, and carbazoles have been detected by these means in our Synthoil work. Workers at the Energy Research and Development Administration/Pittsburgh Energy Research Center (ERDA/PERC) identified a very large number of nitrogen compounds using high-resolution, low-voltage MS in the analysis of chromatographic fractions obtained from Synthoil products (*23*).

In our work NMR has been used mainly to determine structural parameters of the aromatic skeleton. Retcofsky, et al. have used C-13 NMR to determine the C–OH functional groups in phenolic fractions (*24*).

The information summarized above indicates that the heteroaromatic components in coal extracts and coal products possess one-to-eight ring condensed aromatic skeletons, associated with functional groups including furanic, hydroxylic, carboxylic, pyrrolic, pyridinic, thiophenic, and thiolic groups or moieties. Compounds with more than one functional group are present in minor concentrations in the cyclohexane-soluble fractions, but increase significantly in abundance in the cyclohexane-insoluble, benzene-soluble asphaltene fractions. Mass spectrometric evidence indicates the presence of homologous series and proton NMR data indicate that naphthenic rings are also present in the molecules.

Quantitative Analysis. Quantitative analysis of heterocompounds by the high-resolution, low-voltage approach is based on a set of calibration coefficients obtained from pure compounds, extrapolation, and theoretical considerations. Generally, low-voltage sensitivities of heterocompounds increase with aromaticity (number of pi electrons) and decrease with side-chain length. This is analogous to the behavior of aromatic compounds (*25, 26*). A furanic or thiophenic ring is equivalent to an aromatic ring in this sense, thus benzothiophene has almost the same low-voltage sensitivity as naphthalene and dibenzothiophene has almost the same low-voltage sensitivity as phenanthrene. Hydroxyaromatics have significantly higher low-voltage sensitivities than the corresponding aromatics; the sensitivity ratio between phenol and benzene is, for example, 1.96. However, this ratio decreases for the more condensed compounds as the effect of the hydroxy group is "diluted" with respect to a larger molecule.

Pyrrolic-type nitrogen compounds are more sensitive than the corresponding aromatics, while the latter are more sensitive than the pyridinic types. Thus, indole has a higher sensitivity than naphthalene; naphthalene a higher one than quinoline. This effect becomes less pronounced as the aromatic condensation increases (*6*).

All quantitative calculations are carried out by computer programs (*2, 6, 16*). Calibration data are at present available for 95 heteroaromatic compound types, including one-to-seven or -eight ring thiophenes, furans, mono- and dihydroxyaromatics, pyrroles, pyridines, and nitrogen/oxygen compounds. Semiquantitative data are available on an essentially unlimited number of additional heteroaromatics.

In addition to the concentrations of the individual carbon-number homologs, these programs also calculate the following properties:

On Each Compound Type	On a Total Sample or Fraction
Weight percent	Average carbon number
Average carbon number	Average molecular weight
Average molecular weight	Elemental analysis
Elemental analysis	Distillation (only on monosulfur and mono-oxygen compounds)

Data on the nitrogen compounds are normalized to the elemental nitrogen value; all values can be normalized to a given fraction or to a total sample or feed basis. (Examples of quantitative data are given in Table V.)

In addition to the above calculations, our computer programs also separate unresolvable sulfur and hydrocarbon components (10). This step is carried out before calculating the concentrations. It is required because of the inherent difficulties in resolving certain hydrocarbon/sulfur doublets even with an instrumental resolving power of ca. 100,000.

This mathematical separation is based on the fact that the sulfur homologous series start at a lower molecular weight than the interfering hydrocarbon series. For example, the nominal molecular weight of benzothiophene is 134, that of the interfering cyclopentanophenanthrene is 190. Thus there is a region, in this case at molecular-weight values of 134, 148, 162, and 176 at which the entire peak height of the sulfur/hydrocarbon doublet can be assigned unequivocably to benzothiophene and alkylbenzothiophenes. (Other components such as alkylbenzenes, hydroxytetralins, etc. also appear at these nominal molecular weights, but these can be resolved instrumentally without difficulties.) At m/e 190 and above, both alkylbenzothiophenes and cyclopentanophenan-

Table V. Partial Carbon-Number Distribution of

$C No.$	$C_nH_{2n-6}O$	$C_nH_{2n-12}O$	$C_nH_{2n-18}O$
10	0.68	0.02	0.00
11	0.36	0.04	0.00
12	0.13	0.06	0.00
13	0.11	0.04	0.00
14	0.05	0.03	0.02
15	0.04	0.03	0.02
16	0.02	0.07	0.04
17	0.02	0.06	0.07
18	0.02	0.04	0.08
19–28	0.00	0.10	0.29
Total	6.19	0.49	0.52
Av mol wt	126	210	263
Wt % O or N	0.79	0.04	0.03

^a Wt % on sample.

threnes (and/or dihydropyrenes, starting at m/e 204) and their alkyl homologs are possible. Separation of the sulfur/hydrocarbon doublet in this region is based on the carbon-number distribution of the unequivocally assigned sulfur compounds in the 134–176 region. This distribution is extrapolated to the 190$^+$ region using the distribution of a hydrocarbon type that has the same boiling-point characteristics as the sulfur-compound type in question. For benzothiophenes, this hydrocarbon type is naphthalene. Similar procedures are applied to all other sulfur compounds with the calculations carried out for sets of four types at a time to provide changing bases for the extrapolation. The procedure is illustrated in Table VI; the examples given are dibenzothiophenes, nuclear molecular weight 184, and cholanthrenes, nuclear molecular weight 240 ($C_nH_{2n-16}S$ and C_nH_{2n-26} series). The basis of the extrapolation is the carbon-number distribution of the phenanthrene series (C_nH_{2n-18}) whose boiling-point characteristics are similar to those of the dibenzothiophenes.

Applications. The methods summarized above are applicable to coal, petroleum, and shale-derived liquids. Experience acquired during the last 10–15 years shows that they yield reliable information, although obviously there are more difficulties in the analysis of heteroaromatics than in the analysis of aromatic hydrocarbons.

The general validity of the approach is corroborated by the representative data given in Table VII, that show a very good agreement between elemental sulfur, oxygen, and nitrogen values calculated from the MS composition and conventional measurements obtained respectively by x-ray fluorescence (XRF), neutron activation analysis (NAA), and chemical determinations.

Selected Polar Aromatic Types in a Synthoil Product[a]

$C_nH_{2n-22}O$	$C_nH_{2n-16}O_2$	$C_nH_{2n-15}N$	$C_nH_{2n-21}N$
0.00	0.00	0.00	0.00
0.00	0.00	0.00	0.00
0.00	0.00	0.18	0.00
0.00	0.02	0.14	0.00
0.00	0.07	0.14	0.00
0.00	0.08	0.13	0.02
0.01	0.08	0.12	0.04
0.01	0.09	0.11	0.05
0.03	0.07	0.08	0.04
0.27	0.05	0.17	0.19
0.32	0.46	1.07	0.34
289	243	216	266
0.02	0.13	0.07	0.02

Table VI. Mathematical Separation of Unresolved Sulfur
and Hydrocarbon Components[a]

m/e	Total Peak Height	$C_nH_{2n-16}S$	C_nH_{2n-26}	m/e	C_nH_{2n-18} Series Peak Height
		Assigned to			
184	1218	1218	0	178	11236
198	6715	6715	0	192	29456
212	12696	12696	0	206	40128
226	11352	11352	0	220	41423
240	9484	7405	2079	234	27024
254	7348	3592	3756	248	13112
268	5390	2516	2874	262	9184
282	4000	1347	2653	276	4920
:	:	:	:	:	:
408	88	58	30	402	212

[a] Example: $C_nH_{2n-16}S/C_nH_{2n-26}$ series.

Table VII. Elemental Analyses

Sample	Sulfur		Oxygen		Nitrogen	
	MS	XRF	MS	NAA.	MS	Chem
A	0.37	0.29	0.71	0.76	0.90	0.70
B	1.51	1.42	2.40	2.63	5.92	6.00
C	2.61	2.62	3.68	3.22	—	—

Table VIII. Heterocompounds in Selected Synthoil
Feeds and Products

	Wt %		
		Synthoil Product	
	Kentucky Feed Coal	Mildly Hydro-genated	Severely Hydro-genated
Thiophenes in aromatics (MS)	13.35	1.51	0.27
Sulfur in aromatics (XRF)[a]	5.65	0.44	0.09
Furans in aromatics (MS)	9.48	9.66	2.37
Oxygen in aromatics (NAA)	0.75	0.57	0.14

[a] Includes elemental sulfur.

The method is also corroborated by the excellent agreement observed between elemental sulfur level, aromatic thiophenes, and processing conditions in our recent work on the characterization of Synthoil feeds and products (3). A few data excerpted from this effort are given in Table VIII.

Conclusion

High-resolution MS provides significant information on the heterocompounds present in coal liquids. The technique can be applied equally well to coal extracts and to coal liquids. This information can be related to processing parameters. The method supports research and development efforts designed to either liquefy coal or to upgrade coal liquefaction products. The approach is limited only by instrumental resolving power and sample volatility.

Although the method can be applied directly to unseparated samples, the most detailed results are obtained when it is combined with separation methods and spectrometric measurements by IR, NMR, GC/MS, etc. techniques.

Future improvement needs include increased efforts for calibration work, work to extend volatility limits using the direct insertion probe or the field desorption technique (7), and the development of fast and sharp separation methods for specific compound classes.

Acknowledgment

The research discussed in this chapter was supported in part by the U.S. Energy Research and Development Administration, Contract E-(49-18)-2353 and E-(36-1)-8007 and the Electric Power Research Institute, Agreement No. RP 778-1.

Literature Cited

1. Sharkey, A. G.; Shultz, Janet; Friedel, R. A. *Fuel* **1959**, *38*, 315.
2. Aczel, T. *Rev. Anal. Chem.* **1971**, *1*, 226.
3. Aczel, T.; Williams, R. B.; Pancirov, R. J.; Karchmer, J. H. "Chemical Properties of Synthoil Products and Feeds," Report prepared for U. S. Energy Research and Development Administration, FE8007, 1977.
4. Woodward, R. W.; Sturm, G. R.; Vogh, J. W.; Holmes, S. A.; Dooley, J. E. *Proc. Annu. Conf. Mass Spectrometry, 24th, 1976*, p. 487.
5. Shultz, J. L.; White, C. M.; Schweighardt, F. K.; Sharkey, A. G., Jr. *Proc. Annu. Conf. Mass Spectrometry, 25th, 1977*, pp. 145–147.
6. Aczel, T.; Lumpkin, H. E. *Proc. Annu. Conf. Mass Spectrometry, 19th, 1971*, pp. 328–330.
7. Anbar, M.; Kempster, E.; Scolnick, M.; St. John, G. *Proc. Annu. Conf. Mass Spectrometry, 24th, 1976*, pp. 467–468.
8. Lumpkin, H. E.; Taylor, G. R. *Anal. Chem.* **1961**, *33*, 467.

9. Lumpkin, H. E. *Anal. Chem.* **1964**, *36*, 2399.
10. Johnson, B. H.; Aczel, T. *Anal. Chem.* **1967**, *39*, 682.
11. Sheppele, S. E.; Greenwood, G. J.; Grizzle, P. L.; Marriott, T. D.; Perreira, N. B. *Proc. Annu. Conf. Mass Spectrometry, 24th, 1976*, pp. 482–486.
12. Aczel, T.; Lumpkin, H. E. *Proc. Annu. Conf. Mass Spectrometry, 25th, 1977*, pp. 135–138.
13. Lumpkin, H. E.; Wolstenholme, W. A.; Elliott, R. M.; Evans, S.; Hazelby, D. *Proc. Annu. Conf. Mass Spectrometry, 23rd, 1975*, pp. 235–237.
14. Lumpkin, H. E.; Aczel, T., *Proc. Annu. Conf. Mass Spectrometry, 25th, 1977*, pp. 150–152.
15. Aczel, T.; Foster, J. Q.; Karchmer, J. H. "Abstracts of Papers," 157th National Meeting of the American Chemical Society, Minneapolis, MN, April 1969.
16. Aczel, T.; Allan, D. E.; Harding, J. H.; Knipp, E. A. *Anal. Chem.* **1970**, *42*, 341.
17. 1975 Book of ASTM Standards, Method D2007, pp. 163–170.
18. Aczel, T.; Williams, R. B.; Pancirov, R. J.; Karchmer, J. H. "Chemical Properties of Synthoil Products and Feeds," pp. 255–259, Report prepared for U.S. Energy Research and Development Administration, FE8007, 1977.
19. Sternberg, H. W.; Raymond, R.; Schweighardt, F. K. *Science* **1975**, *188*, 49.
20. Shultz, J. L., ERDA/PERC, personal communication.
21. Farcasiu, M. *Fuel* **1977**, *56*, 9–14.
22. Ruberto, R. G.; Jewell, D. M.; Jensen, R. K.; Cronauer, D. C. *Am. Chem. Soc., Div. Fuel Chem., Prepr.* **1974**, 258.
23. Schultz, J. L.; White, C. M.; Schweighardt, F. K.; Sharkey, A. G., Jr. PERC/RI-77/7.
24. Retcofsky, H. L.; Link, T. A., Abstracts of the 1977 Pittsburgh Conference on Analytical Chemistry, 1977, No. 13.
25. Lumpkin, H. E. *Anal. Chem.* **1958**, *30*, 321.
26. Lumpkin, H. E.; Aczel, T. *Anal. Chem.* **1964**, *36*, 181.

RECEIVED June 28, 1978.

Catalytic Hydroprocessing of Shale Oil to Produce Distillate Fuels

RICHARD F. SULLIVAN and BRUCE E. STANGELAND

Chevron Research Company, Richmond, CA 94802

Paraho shale oil is too high in nitrogen (2.2%) and other contaminants to be used directly as a fuel in most current applications. The key to successful shale oil refining is a hydrotreating process to remove these contaminants. In this study, nitrogen was reduced to concentrations as low as 1 ppm in the hydrotreated whole oil in a single catalytic stage using a catalyst containing nickel, tungsten, silica, and alumina. However, it is economically preferable to hydro-treat at less severe conditions to convert the shale oil to a premium synthetic crude containing about 500 ppm of nitrogen. This synthetic crude resembles a typical hydrorefined petroleum gas oil and is suitable for downstream processing to specification transportation fuels in conventional modern refineries.

This chapter presents results of a Chevron Research Company study sponsored by the U.S. Department of Energy (DOE) to demonstrate the feasibility of converting whole shale oil to a synthetic crude resembling a typical petroleum distillate. The synthetic crude thus produced then can be processed, in conventional petroleum-refining facilities, to transportation fuels such as high-octane gasoline and diesel and jet fuels. The raw shale oil feed used in this study is a typical Colorado shale oil produced in a surface retort in the so-called indirectly heated mode.

Crude shale oil is an important potential source of transportation fuels when properly refined. However, although it is low in sulfur compared with mid-East crudes, it is much higher in nitrogen than typical petroleum crudes. The shale oil used in this study contains 2.2 wt % nitrogen; typical petroleum crudes contain less than 0.3% nitrogen.

0-8412-0456-X/79/33-179-025$06.75/1

Nearly all of this nitrogen must be removed from the shale oil before it is suitable for use as a transportation fuel.

Nitrogen can act as a poison for both hydroprocessing catalysts and fluid catalytic cracking (FCC) catalysts. Therefore, its presence complicates the conversion of shale oil to finished products.

In this study, whole shale oil is catalytically hydrodenitrified to reduce the nitrogen to levels as low as 1 ppm in a single catalytic stage. However, for economic reasons, it appears preferable to denitrify to about 0.05 wt % nitrogen. The resulting synthetic crude resembles a petroleum distillate that can be fractionated and further processed as necessary in conventional petroleum-refining facilities.

Shale oil contains about 0.6% sulfur. Sulfur is removed more easily by hydrotreating than nitrogen; therefore, only a few parts per million of sulfur remain at a product nitrogen of 0.05 wt %. Oxygen contained in the shale oil is reduced also to low levels during hydrodenitrification. The shale oil contains appreciable quantities of iron and arsenic which are also potential catalyst poisons. In this study, these metals are removed by a guard bed placed upstream from the hydrotreating catalyst.

The feasibility of hydrotreating whole shale oil is demonstrated by means of several long pilot plant tests using proprietary commercial catalysts developed by Chevron. One such test was on stream for over 3500 hr. The rate of catalyst deactivation was very low at processing conditions of 0.6 LHSV and 2000 psia hydrogen pressure. The run was shut down when the feed supply was exhausted although the catalyst was still active.

Hydrotreated shale oil has an advantage as a refinery feed. In contrast to most petroleum crude oils, it contains essentially no residuum. Properties of the hydrotreated product from whole shale oil are similar to those of distillate fractions from waxy petroleum Arabian or Sumatran crudes. An exception is the sulfur content which is much lower for hydrotreated shale oil than for most crudes.

The 650°F+ fraction of the hydrotreated shale oil was processed in in FCC pilot plant, and the results show that it is an excellent feed for a conventional refinery. The resulting FCC yields, activity, and product qualities are quite similar to those derived from normal petroleum gas oils.

The properties of the 650°F+ fraction of the hydrotreated shale oil also indicate that it will be an appropriate feed to an extinction recycle hydrocracker in situations where jet fuel is a desired product. Supporting pilot plant studies are described elsewhere (1, 2, 3).

Based on correlations, the naphthas from the shale oil hydrotreater can be readily upgraded to high-octane gasolines by catalytic reforming. The middle distillate fractions will require some additional hydrotreating to produce salable diesel or jet fuel. The technology is available, and results of pilot plant studies are reported in a DOE report (*1*) and other papers (*2, 3*).

The results of this study are, therefore, quite encouraging. Shale oil appears to be an attractive feed for processing in existing refineries that have appropriate hydrotreating facilities.

Background

The oil shale deposits of Colorado, Utah, and Wyoming contain a vast reserve of hydrocarbons which could supply a significant fraction of our energy needs. The deposits of the known shale oil resources in the Green River Formation are estimated to contain the equivalent of 2,000,000,000 bbl of crude oil (*4*). However, as produced from a retort, shale oil is too high in nitrogen, sulfur, and arsenic content to allow direct utilization as a fuel in most current applications. The goals of many research groups over the years have been to reduce the amounts of these contaminants and crack the shale oil into lighter products (*5–22*). Many of the earlier reports have concentrated on the hydroprocessing of either the raw shale oil, distillate fractions from the raw shale oil, or coker distillates from shale oil and have not tied together the various process units that would be needed in a complete refinery.

Most of the studies that have been made on shale oil have been done in pilot scale equipment. However, in 1975 a 10,000-bbl test run was made in the Gary Western Refinery near Grand Junction, Colorado (*8*). The processing scheme involved coking Paraho shale oil, hydrotreating the light products, and acid treating the diesel. Pilot plant studies were made (*9*) to determine the appropriate hydrotreater operating conditions and to obtain information on product quality. The products met many of the required specifications but were poor in thermal stability and were quite gummy. Their high nitrogen and oxygen contents could well have caused such problems. More severe hydrotreating should produce better products (*5*).

In a 1977 test run in Chevron U.S.A.'s Salt Lake Refinery, hydrotreating facilities were adequate; stable products were produced by coking a mixture of in situ shale oil and crude oil residua (*3*).

In another coking study on shale oil, light gas oil product was acid and caustic treated to reduce the nitrogen from 1.71 wt % to 0.05 wt %. A commercial railroad diesel engine ran for 750 hr on this fuel with no unusual deposits, accumulations, or problems with poor performance (*5*). Cottingham and Nickerson (*6*) report that hydrotreating raw, in situ shale oil, without coking, gives high yields of diesel fuel (about 52%).

Thermal stability of this fuel was not reported, although it did meet other specifications.

Making salable jet fuels is expected to be more difficult than making diesel fuel because of the more stringent stability specifications. One approach is to distill a kerosene cut from raw shale oil and to hydrotreat it separately. Kalfadelis (15) reports that some problems were encountered in trying to lower nitrogen below 40 ppm, which is still a much higher nitrogen content than typical of conventional jet products. However, this processing route has produced a jet fuel that passed the JFTOT Thermal Stability Test (10).

An additional problem in processing of shale oil compared with petroleum processing is the presence of appreciable quantities of arsenic in shale oil. Arsenic can be a poison for hydrodenitrification catalysts (19). Various pretreatment steps, such as contacting with sodium hydroxide (19) or using a catalytic guard bed (20) or noncatalytic heat treatment (21), have been proposed to remove the arsenic prior to hydrodenitrification.

Although the oxygen content of 1.16% for the shale oil is higher than most crudes, some California crudes have similar oxygen contents. Therefore, unusual processing problems are not expected as a result of the presence of this oxygen in shale oil.

The objectives of the DOE-sponsored work described here are to determine the feasibility and estimate the economics of converting shale oil to transportation fuels using presently available petroleum-refining technology. This conversion is based on modern catalytic hydroprocessing techniques developed by Chevron which convert the shale oil to a synthetic crude oil that resembles a typical petroleum distillate. The synthetic crude thus produced then can be processed in conventional petroleum-refining facilities to make high-octane gasoline and jet and diesel fuels.

More detailed results are given in a DOE report (1). The program also will include hydroprocessing of three coal-derived liquids. Additional results will be reported in future works.

Pretreatment of Raw Shale Oil

The raw Colorado shale oil used in this study was provided by DOE's Laramie Energy Research Center (LERC). It was prepared in the Paraho tests in a surface retort operated in the so-called indirectly heated mode. As received, it was an emulsion containing 6% water and about 0.5% fine particles in the shale oil. Heating this mixture to 170°F broke the emulsion; most of the water and fines settled out after standing at

170°F for 6 hr. The dewatered shale oil was filtered through a 15-μ filter; however, filtration did not take out appreciable amounts of additional fines.

Inspections of the dewatered whole shale oil are given in Table I.

Table I. Properties of Dewatered Paraho Shale Oil[a]

Pilot Plant Feed

Gravity (°API)	20.2
Sulfur (wt %)	0.66
Total nitrogen (wt %)	2.18
Oxygen (wt %)	1.16
Arsenic (ppm)	28
Pour point (ASTM, °F)	90
Carbon (wt %)	84.30
Hydrogen (wt %)	11.29
Hydrogen/carbon atom ratio	1.60
Chloride (ppm)	< 0.2
Ash (wt %) (ASTM D 486)	0.03
Iron (ppm)[b]	70
Filter residue ash (0.45 μ-filter)	
total solids (ppm)	252
ash (ppm)	194
sediment (plus trace water) (vol %)	0.1
bromine no.	51
average molecular weight	326
Viscosity (cSt)	
122°F	25.45
210°F	5.54
acid neutralization no. (mg KOH/g)	2.3
base neutralization no. (mg KOH/g)	38
(equivalent)	
pH	9.2
maleic anhydride no. (mg/g)	40.6
navy heater test	No. 1 rating
hot heptane asphaltenes (including any fines) (wt %)	0.17
ASTM D 1160 distillation (°F) (At 10 mm)	
St/5	386/456
10/30	508/659
50	776
70/90	871/995
EP	/1022 (94%)
% overhead (excl. trap)	94
% in trap	1
% in flask	5

[a] Produced by indirectly heated mode.
[b] This iron is not removed by filtration through a 0.45-μ filter.

Table II.   Properties of

Boiling Range (°F)	Gravity (°API)	Sulfur (wt %)	Total Nitrogen (wt %)	Arsenic (ppm)
266–400	38.3	1.08	1.55	6
400–650	29.4	0.74	1.58	38
650–800	20.5	0.66	2.14	21
800–835	18.6	0.54	2.07	18
835+[a]	10.4	0.52	2.53	35

[a] Other properties of the 835°F+ bottoms are as follows: n-C_5 asphaltenes, 8.35 wt % ; hot C_7 asphaltenes, 5.41 wt % ; mol wt, 1227; Ramsbottom carbon, 13.75%. (Note: fines are included as asphaltenes in these analyses.)

Properties are characteristic of typical surface-retorted Colorado shale oils (5, 7, 11, 12, 13). Table II gives properties of fractions of the shale oil from a small-scale batch distillation.

When the filtered, dewatered oil is ashed via standard ASTM tests, x-ray analysis of the ash shows the presence of $Ca_2Al_2SiO_7$, $CaMgSi_2O_7$, and Fe_3O_4.

Whole Oil Hydrotreating

Removal of nitrogen is probably the single most important step in processing raw shale oil, because nitrogen is a poison for cracking catalysts (13, 17, 18). Various workers (10, 15) have fractionated the shale oil and hydrotreated the distillate fractions to reduce the nitrogen content. Shaw et al. (14) suggested that a reasonable processing route for shale oil is to hydrotreat the whole oil to reduce its nitrogen content and then to process at least the 900°F⁻ material in a conventional refinery.

Our approach has been to hydrotreat the whole shale oil before distillation; in this way, the entire stock is upgraded in this first catalytic step. Results support the conclusion that hydrodenitrification of whole shale oil is commercially feasible.

Screening runs were made for several catalysts and at product nitrogen concentrations as low as 1 ppm. Evaluation of the results by the Chevron Research Process Engineering Department indicated that it is economically advantageous to hydrotreat to about 500–1500 ppm nitrogen in the whole liquid product, then to distill, and, as necessary, process the various product fractions. The Chevron catalyst of choice for whole oil hydrotreating was determined to be ICR 106, based on results of the catalyst screening tests. This catalyst contained nickel, tungsten, silica, and alumina.

Fractions from Paraho Shale Oil

Oxygen (wt %)	Bromine No.	Pour Point (ASTM, °F)	Viscosity (cSt)		
			122°F	210°F	300°F
1.12	79	10	0.905		
1.26	60	80	3.241		
0.97	53	110	19.64	4.49	
0.78	46	95	63	9.04	
1.24			2966	97.73	15.63

Hydrotreating is done in fixed-bed, downflow pilot plants. They are equipped with high- and low-pressure product separators, recycle hydrogen facilities, product distillation columns, and extensive control and monitoring instruments. Catalyst is sulfided in situ. These units generally give good recovery of products (within 2%) and provide a measure of hydrogen consumption.

The first run with ICR 106 catalyst (Run 81–4) was a 2000-hr test made in a relatively small pilot plant containing 130 mL of catalyst to determine denitrification kinetics and hydrogen consumption. Activity data are plotted in Figure 1 for 0.2, 0.3, and 0.6 LHSV. A simple first-

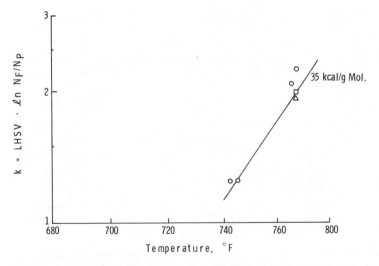

Figure 1. Denitrogenation kinetics of whole shale oil on ICR 106: (○), 0.6 LHSV; (□), 0.3 LHSV; and (△), 0.2 LHSV. $P_{H_2} = 1750$–1850 psia.

order kinetic expression with an activation energy of 35 kcal/g mol gives a reasonable fit to the data as has been found to be typical for petroleum stocks with this catalyst. Product nitrogen ranged from 2800 ppm down to 1.3 ppm. Hydrogen consumptions from these tests are shown in Figure 2. At 500–1000-ppm nitrogen levels, the hydrogen consumption was ca. 2000 scf/bbl. This is in good agreement with the results of Frost and co-workers (13, 16) who report consumptions of 1800–2000 scf/bbl at similar product nitrogen levels. Inspections and yields obtained during Run 81–4 are given in Table XI.

This run demonstrates the feasibility of reducing nitrogen from 21,800 ppm to about 1 ppm in a single reactor stage. The low feed rates required, however, suggest that a more economic operating point is in the 500–1500-ppm nitrogen range. At this product nitrogen concentration the 650°F+ product is similar to normal gas oil feeds to existing hydrocrackers and FCC units. Table XI shows that some cracking of shale oil occurs during denitrification, and that the amount of 1000°F+ material is reduced to about 1% of the whole product at a nitrogen concentration of 500 ppm. This end point reduction is desirable for downstream conversion processes.

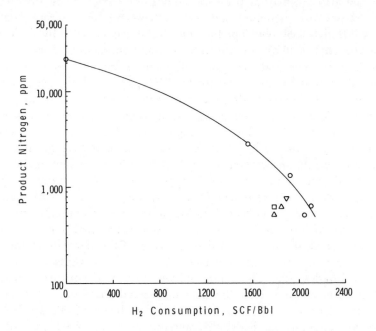

Figure 2.   Hydrogen consumption vs. product nitrogen in the hydrotreating of whole shale oil with ICR 106. Small pilot plant: (○), about 1850 psia. Large pilot plant: (□), about 1550 psia; (△), approximately 1800 psia; and (▽), about 1650 psia.

This first run with ICR 106 catalyst was shut down at 2000 hr on stream when a plug consisting of iron and arsenic deposits from the shale oil developed in the preheat section of the reactor. (In this first test, a guard bed to remove these metals was not provided.) To prepare sufficient feedstock for downstream processing studies, a larger-scale, 3500-hr pilot plant run was made with 650 mL of ICR 106 catalyst.

In this second run, alumina of moderate surface area was used as a guard bed to remove arsenic and iron upstream from the catalyst. The guard bed was designed to protect the downstream catalyst from these poisons. Cottingham and Carpenter (*12*) found that fairly mild conditions of 650°F and 1000 psig are effective in the presence of a hydrogenation catalyst for removing substantial amounts of these metals. Curtin (*21*) has disclosed that noncatalytic thermal treatment is effective for removing arsenic and selenium from synthetic crudes at 750°–850°F in the presence of hydrogen at 500–1500 psig. Our results show that the guard bed not only removes the arsenic and iron but also traces of zinc and selenium that are present in the shale oil.

The time history of this run (86–51) is shown in Figure 3. After a 300-hr line-out period, the product nitrogen settled in the 500–1000-ppm range. Hydrogen partial pressure slowly drifted down as the catalyst temperature was increased and the methane production rate rose. At about 960 hr, a leak developed in the recycle gas system. By the time it was corrected, at 1200 hr, the rate of nitrogen removal had increased substantially. The leak had bled off enough methane to increase the hydrogen partial pressure from 1650 to 2100 psia. With the leak repaired, hydrogen partial pressure dropped again and the reaction rate decreased. Because of the dramatic effect of hydrogen partial pressure on reaction rate, an intentional gas bleed of about 500 scf/bbl was introduced at 1440 hr. By 1510 hr, hydrogen pressure had increased by about 300 psia and the rate of nitrogen removal showed improvement again.

At this point, our original shipment of raw shale oil ran out. While waiting for the next shipment from Laramie, we reran some partially hydrotreated shale oil collected during off-test periods on other runs.

Shortly before the feed ran out, it was noted that a pressure drop had developed across the guard bed owing to the deposits of iron and arsenic. By the time the new feed was ready, the pressure drop had increased to 800 psia. The guard bed was recharged with fresh material, and the run continued on the new dewatered shale oil. This oil was virtually identical to the first one. A recycle bleed of 500 scf/bbl was used to maintain the hydrogen pressure at about 2000 psi. At this pressure, the catalyst deactivation rate was low; the catalyst fouled at an average rate of less than 0.01°F/hr during the last 2000 hr of the run. The guard bed was replaced for a second time at 3260 hr on stream.

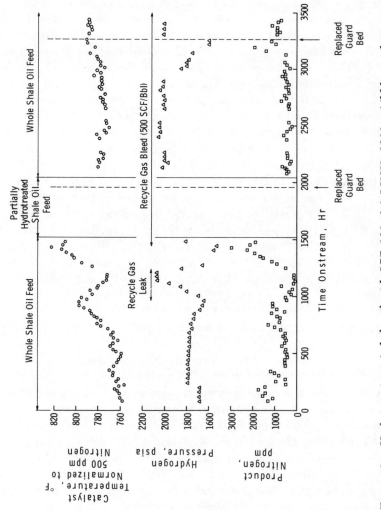

Figure 3. Hydrotreating of shale oil with ICR 106: 0.6 LHSV; 8000 scf/bbl recycle gas rate.

Table III.   Properties of Narrow Boiling Mid-Distillate
Fractions from Hydrotreated Shale Oil Blend

	Fraction No.				
	1	2	3	4	5
Boiling range (°F)	400–450	450–500	500–550	550–600	600–650
Yield (LV % of whole hydrotreated shale oil blend)	8.8	8.7	8.3	9.5	11.8
Gravity (°API)	41.1	38.4	36.1	35.7	33.2
Aniline point (°F)	131.1	140.5	147.7	162.9	170.3
Nitrogen (ppm)	485	610	790	735	990
Freeze point (°F)	−43	−11	+1	+25	+49
Pour point (°F)	−60	−20	−5	+20	+45
Cloud point (°F)	−54	−20	−4	+20	+45
Group type (LV %)					
paraffins	28.5	32.6	32.1		
naphthenes	45.7	40.8	44.1		
aromatics	25.8	26.6	23.9		

Whole liquid product from this run was blended and batch distilled
to make feed for downstream processing studies. Only products containing less than 1000 ppm of nitrogen and generated during the first 1500 hr
of the run were used for the tests described in this chapter. Inspections
on some of the narrow boiling cuts of this product are given in Table III.
The rapid rise in freeze and pour points with an increasing boiling point
will determine the allowable end point for distillate fuels. High wax
contents have been noticed in the past (8, 9) in shale oil products. The
350°–650°F diesel cut is similar in inspections to diesel fuels made by
acid treating (5) and hydrocracking (6) shale oil, as shown in Table IV.

Table IV.   Comparison of Diesel Fuels from Processed Shale Oil

	Chevron Research Hydrotreated Shale Oil	Acid-Treated Diesel (5)	In Situ Hydro-cracked Truck–Tractor Diesel (6)
Gravity (°API)	37.4	35.0	39.7
Flash point (°F)	175 est.	176	212
Pour point (°F)	0	+15	−30
Cetane no.	52 est.	46	54
Sulfur (wt %)	0.0004	0.90	0.02
Nitrogen (wt %)	0.054	0.05	0.0166
Distillation (°F)			
10%	437	454	442
50%	508	506	484
90%	586	607	605
Viscosity (cSt at 100°F)	2.82		2.40

Table V. Properties of Naphtha Fractions from Hydrotreated Shale Oil Blends and from Arabian Light Crude

	Naphtha from Hydrotreated Shale Oil		*Raw Naphtha from Arabian Light Crude*
Boiling range (°F)	C_5–180	180–350	180–350
Inspections			
gravity (°API)	83.9	55.5	59.2
aniline point (°F)		127.3	135.8
sulfur (ppm)	7.3	2.7	304
nitrogen (ppm)	1.8	56	< 2
Octane no.			
F-1 clear		46.0	32.8
Group type (LV %)			
paraffins	88.5	50.8	67.9
naphthenes	10.8	38.3	21.6
aromatics	0.6	10.8	10.5
olefins	0.1		
ASTM D 86 distillation (°F)			
St/5		188/220	208/225
10/30		231/259	231/248
50		279	267
70/90		299/320	292/321
95/EP		328/362	339/377
% Overhead (incl. trap)		99	99

Table VI. Comparison of Straight-Run Vacuum Gas Oils (VGO) with 650°F+ Hydrotreated Shale Oil

	Arabian VGO	*Hydrotreated Shale Oil*		*Sumatran VGO*
Inspections				
gravity (°API)	22.2	30.9	30.8	31.9
aniline point (°F)	183	212	210	217
sulfur	2.46 wt %	< 10 ppm	< 10 ppm	1300 ppm
nitrogen (ppm)	780	385	870	385
oxygen (ppm)	1100	150	200	1000
pour point (°F)	100	105	100	110
Group type (LV %)				
paraffins	18		21	44
naphthenes	22		46	31
aromatics	33		33	23
sulfur comp.	27		0	2
Hot heptane asphaltenes (ppm)	120	477	756	132
ASTM D 1160 distillation (°F)				
St/5	707/738		550/716	550/640
10/30	752/795		729/756	659/756
50	858		802	809
70/90	917/978		842/916	859/940
95/EP	998/1038		944/1073	983/1019

The 180°–350°F heavy naphtha is compared in Table V with a raw naphtha derived from Arabian light crude oil. The shale oil naphtha is a better reformer feed than the petroleum stock: it is higher in octane and naphthenes and lower in paraffins.

Inspections of the 650°F+ shale oil at two different nitrogen levels are compared in Table VI with those of typical feeds to refinery cracking units: raw vacuum gas oils from Arabian Light and Sumatran crude oils. Generally, the shale oil properties are intermediate. The hydrotreated shale oils are much lower in sulfur and oxygen but are similar in nitrogen and pour point. Thus, they should be good feeds to conventional downstream refining units.

FCC Processing

The FCC process is the most common conversion unit in use today. As such, it is important to determine the performance of an FCC when feeding hydrotreated shale oil. The two 650°F+ feeds shown in Table VI were evaluated in an FCC pilot plant operating in a fixed fluidized-bed mode. The catalyst was withdrawn from an operating commercial FCC unit. It is a zeolite catalyst, CBZ-1, produced by Davison Chemical Division of W. R. Grace and Company and is moderately active as well as contaminated with metals.

The fixed fluidized-bed test unit is operated in a cyclic manner. Preheated feed enters the bottom of a conical-shaped reactor where it vaporizes and fluidizes the catalyst in the reactor. Steam is added with the feed to aid in its vaporization. The reactor is at 975°F during the reaction. Oil is fed for a 5-min period after which a nitrogen purge sweeps the hydrocarbon vapors from the reactor. During this purge, the temperature of the reactor is raised to 1050°F; and with completion of the purge, air is introduced into the reactor to burn the accumulated coke from the catalyst. At the end of the regeneration period, a second nitrogen purge removes the air from the reactor, thus completing the cracking–regeneration cycle. Multiple cycles usually are made to provide sufficient liquid product for distillation and inspection of the distillate fractions.

The hydrocarbon gas from each cycle is collected in a gas holder, mixed, and sampled. Gas composition is determined chromatographically. The coke deposited on the catalyst is determined for each cycle by passing an aliquot of the regeneration gases over hot cupric oxide to convert carbon monoxide to carbon dioxide and then through an Ascarite absorption bulb.

Results on the two hydrotreated feeds and also on one previously tested raw shale oil sample are shown in Table VII. The high nitrogen content of the raw shale oil (1.85 wt %) drastically reduces conversion and increases coke make. The two hydrotreated feeds give high conver-

Table VII. The Effect of Feed Nitrogen on FCC Conversion at 975°F Reactor Temperature

	Raw Shale Oil	Hydrotreated Shale Oil	
Nitrogen (ppm)	18,500	870	385
Conversion (LV %) below 430°F	38.9	75.4	81.0
Coke yield (wt %)	6.21	3.87	4.41

sions and low coke makes and would be hard to distinguish from conventional FCC feeds. Previous work also found hydrotreated shale oils similar to petroleum feeds (11). Table VIII gives a detailed breakdown of product yields for the two hydrotreated feeds. These yields are compared in Figure 4 with those obtained previously on hydrotreated Arabian Light vacuum gas oils (23). The earlier data were generated in

Table VIII. Catalytic Cracking of Hydrotreated Shale Oil

Nitrogen content of feed (ppm)	385		870	
	Wt %	LV %	Wt %	LV %
Conversion	78.84	80.98	73.45	75.38
Product yields				
coke	4.41		3.87	
H_2	0.06		0.07	
C_1	1.12		1.07	
ethane	0.88		0.89	
ethene	0.78		0.71	
Total C_2's	1.66		1.60	
propane	3.17	5.43	2.46	4.23
propene	5.97	9.97	5.15	8.60
Total C_3's	9.14	15.40	7.61	12.83
i–C_4	7.15	11.06	5.73	8.88
n–C_4	2.10	3.14	1.69	2.52
C_4 Olefins	6.45	9.21	6.00	8.57
Total C_4's	15.70	23.41	13.42	19.97
Light Gasoline (C_5—250°F)	29.64	38.35	28.70	37.05
Heavy Gasoline (250°–430°F)	17.11	17.81	17.11	17.97
Total Gasoline (C_5—430°F)	46.75	56.16	45.81	55.02
Light Cycle Oil (430°–625°F)	9.68	8.88	10.73	10.04
Heavy Cycle Oil (625°F+)	11.48	10.14	15.82	14.58
Total Cycle Oil (430°F+)	21.16	19.02	26.55	24.62
C_3+ Liquid Product	92.75	113.99	93.39	112.44

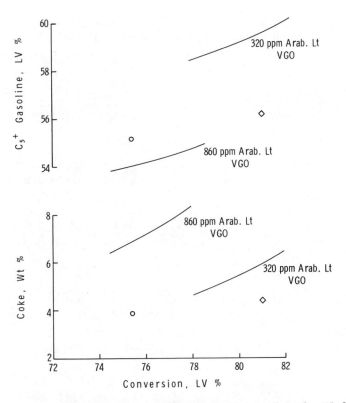

Figure 4. Catalytic cracking yields from hydrotreated feeds. Shale oil and Arabian Light vacuum gas oil at 975°F: (○), shale oil containing 870 ppm N; (◇), shale oil containing 385 ppm N.

a continuous-flow pilot plant. Comparison of results for petroleum stocks from the two units indicate that results should be comparable. For a given conversion and feed nitrogen level, the shale oil feeds produced less coke. At the lower nitrogen level, the shale oil feed produced less gasoline and a correspondingly higher amount of C_3–C_4 product.

Octane numbers of the gasolines for the hydrotreated shale oil and Arabian Light vacuum gas oil are compared in Table IX. Again, the results are similar. At the higher nitrogen level, the octanes are virtually identical on the shale and vacuum gas oil feeds. At the lower nitrogen level, the shale oil feed produced slightly lower octane on the light gasoline and somewhat higher octane on the heavy gasoline. Inspections on all products from the two shale oil feeds are given in Table XII. Although not reported, sulfur contents of all products must be very low; the feed sulfur content of the hydrotreated shale oil is less than 10 ppm.

Table IX. Catalytic Cracking of Hydrotreated Oils; Gasoline Octane Numbers

Feed	Hydrotreated Shale Oil		Hydrotreated Arabian Light Gas Oil			
Feed nitrogen (ppm)	385	870	320		860	
Conversion (LV %)	80.98	75.38	78.12	81.93	74.61	77.84
Octane numbers						
Light gasoline (C_5— 250°F)						
F-1 clear	89.0	90.3	91.2	—	90.2	—
F-2 clear	79.1	79.5	80.8	—	78.8	79.4
Heavy gasoline (250°– 430°F)						
F-1 clear	95.3	93.3	94.4	94.4	93.0	93.6
F-2 clear	84.6	82.8	83.3	84.2	82.8	83.5

Shale Oil Refinery

One of the aims of this research study is to assess the economic feasibility of shale oil hydrotreating. These cost studies are reported elsewhere (1, 3).

A flow diagram for a proposed shale oil refinery is shown in Figure 5. An FCC unit is used as the primary cracking process. This refinery produces high-octane gasoline and diesel fuel. Jet fuel also could be produced by severely hydrogenating a kerosene cut (10) from the whole-oil hydrotreater or by using a hydrocracker in place of the FCC. The

Table X. Pilot Plant Data on Whole Oil Hydrotreating with ICR 106

Processing Conditions	765°F, 0.6 LHSV, 1800 psia H_2, 1850 SCF/B H_2 Consumption		
Yields, on unit fresh feed basis	Gas—2.0 Wt %; Total C_5+ Liquid—106.9 LV %		
LV % yield to raw shale oil	14.7 To Cat. Ref.	49.3 Mid. Dist.	41.0 FCC Feed
Product inspections			
Boiling range (°F)	180–350	350–650	650+
Gravity (°API)	55.5	37.4	30.8
Aniline point (°F)	127	147	210
Nitrogen (ppm)	56	540	870
Pour point (°F)		0	+100
Viscosity (cSt at 122°F)		2.5	16.4
Octane number			
F-1 clear	46.0		
F-2 clear			

mid-distillate hydrotreater probably will be needed to reduce diesel nitrogen enough to make stable product. Heavy naphtha from the whole-oil hydrotreater is denitrified in the first stage of a catalytic reformer and then reformed over a bimetallic catalyst. Gasoline is a blend of this reformate, FCC light and heavy gasoline, a C_5–C_6 cut from the hydrotreater, and butanes. The diesel fuel could contain some light-cycle oil from the FCC. Refinery fuel would consist of light gases, heavy-cycle oil, and perhaps some light-cycle oil.

Approximate yields for the hydrotreating and FCC processes are shown in Table X. The whole-oil hydrotreater data are based on results from the large-scale feed preparation run. The FCC data are described above. Since 41 LV % of the raw shale oil is fed to the FCC as hydrotreated 650°F+ bottoms, the FCC pilot plant yields were multiplied by 0.41 and reported as yield to raw shale oil.

Both the mid-distillate and the light-cycle oil have low pour points, but their nitrogen levels ca. 400 ppm may have to be reduced further. Sulfur contents on all products are less than 0.01 wt % because all of the feed passes through the whole-oil hydrotreater. Net diesel yield from such a refinery would be about 50 LV %, and gasoline yield would approach 40 LV %. Detailed yields for a similar refinery operating with FCC feed containing somewhat more nitrogen are given in Refs. *1*, *2*, and *3*. A hydrotreating/FCC refinery is compared with a proposed refinery in which hydrocracking is the major conversion process. It also is compared with one in which the shale oil is coked prior to hydrotreatment.

Processing Paraho Shale Oil

FCC of Hydrotreated 650°F+ Oil with Equilibrium CBZ-1

975°F, 75.4 LV % Conversion

Coke—3.9 Wt %; Gas—23.8 Wt %; Total C_5+ Liquid—79.6 LV %

22.6 *Gasoline*	*4.1* *LCO*	*6.0* *HCO*
C_5—430	430–625	625+
	20.3	18.3
< 80	< 30	
53	388	1400
	—20	+85
	2.4	13
91.3		
80.6		

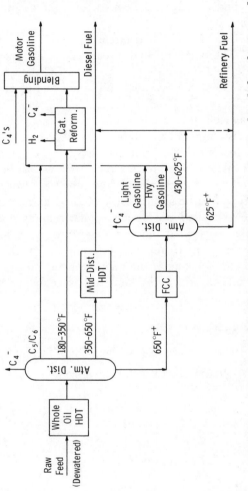

Figure 5. Flow diagram of proposed hydrotreating/FCC refinery for Paraho shale oil. C_4^- is butanes and lighter; use for H_2 plant feed, gasoline blending, and refinery fuel. Foul gas and water are treated to recover NH_3 and sulfur. Low-pressure catalytic reformer uses bimetallic catalyst.

Acknowledgment

The authors wish to thank J. W. Scott for his helpful suggestions throughout this program, C. E. Rudy for providing FCC data, and D. R. Cash and H. A. Frumkin for help in planning pilot plant runs. This chapter was prepared as an account of work performed by Chevron Research Company sponsored by the U.S. Government. Neither Chevron, nor the United States, the U.S. DOE, or any of their employees, contractors or subcontractors, makes any warranty, express or implied, or assumes any legal liability of responsibility for the accuracy, completeness, or usefulness of any information, apparatus, product, or process disclosed, or represents that its use would not infringe privately owned rights. This work was prepared for the U.S. DOE under Contract No. EF-76-C-01-2315.

Literature Cited

1. Sullivan, R. F.; Stangeland, B. E.; Green, D. C.; Rudy, C. E.; Frumkin, H. A. "Refining and Upgrading of Synfuels from Coal and Oil Shales by Advanced Catalytic Processes, First Interim Report, Processing of Paraho Shale Oil," April 1978, U.S. Department of Energy Report FE-2315-25.
2. Sullivan, R. F.; Stangeland, B. E. "Converting Green River Shale Oil to Transportation Fuels," Eleventh Oil Shale Symposium Proceedings, April 12–14, **1978**, Colorado School of Mines Press, p. 120.
3. Sullivan, R. F.; Stangeland, B. E.; Frumkin, H. A.; Samuel, C. W. "Refining Shale Oil," 1978 Proceedings—Refining Department, 43rd Midyear Meeting, May 8–10, **1978**, American Petroleum Institute, 57, 199.
4. Yen, T. F. "Science and Technology of Oil Shale"; Ann Arbor Science Publishers, Inc.: Ann Arbor, Michigan, 1976; Chapter 1.
5. Cottingham, P. L. "Diesel Fuels from Shale Oil," Symposium on Alternate Fuel Resources, Santa Maria, California, March 25, **1976**; CONF 760342-3.
6. Cottingham, P. L.; Nickerson, L. G. "Diesel and Burner Fuels from Hydrocracking In Situ Shale Oil," *ACS Symp. Ser.* **1975**, *20*, 99.
7. Smith, W. M.; Landrum, T. C.; Phillips, G. E. "Hydrogenation of Shale Oil," *Ind. Eng. Chem.* **1952**, *44*, 586.
8. Bartick, H. A.; Hertzberg, R. H.; Sargent, W. H.; Woodcock, B. A. "Alternate Fuels for the Department of Defense," 82nd National Meeting, AIChE, Atlantic City, New Jersey, August 29–September 1, **1976**.
9. Kunchal, S. K. "Coking and Hydrotreating Studies of Paraho Shale Oil," 82nd National Meeting, AIChE, Atlantic City, New Jersey, August 29–September 1, **1976**.
10. Kalfadelis, C. D.; Shaw, H. "A Pilot Plant Study of Jet Fuel Production from Coal and Shale-Derived Oil," 82nd National Meeting, AIChE, Atlantic City, New Jersey, August 29–September 1, **1976**.
11. Frost, C. M.; Carpenter, H. C.; Hopkins, C. B. Jr.; Tihen, S. S.; Cottingham, P. L. "Hydrogenating Shale Oil and Catalytic Cracking of Hydrogenated Stock," **1960**, U.S. Department of Interior, Bureau of Mines Report 5574.
12. Cottingham, P. L.; Carpenter, H. C. "Hydrocracking Prehydrogenated Shale Oil," *Ind. Eng. Chem., Process Des. Dev.* **1967**, *6*, 212.

13. Frost, C. M.; Poulson, R. E.; Jensen, H. B. "Hydrodenitrification of Crude Shale Oil," July 1, 1975, U.S. Energy Research and Development Administration Report LERC/RI 7513.
14. Shaw, H.; Kalfadelis, C. D.; Jahnig, C. E. "Evaluation of Methods to Produce Aviation Turbine Fuels from Synthetic Crude Oils—Phase 1," March 1975, Air Force Aero Propulsion Lab Report AFAPL-TR-75-10.
15. Kalfadelis, C. D. "Evaluation of Methods to Produce Aviation Turbine Fuels from Synthetic Crude Oils—Phase 2," May 1976, Air Force Aero Propulsion Lab Report AFAPL-TR-75-10, Vol. 11.
16. Frost, C. M.; Cottingham, P. L. "Hydrogenating Shale Oil at Low Space Velocity," 1973, U.S. Bureau of Mines Report RI 7738.
17. Scott, J. W.; Bridge, A. G. "The Continuing Development of Hydrocracking," Adv. Chem. Ser. 1971, 103, 113.
18. Silver, H. F.; Wang, N. H.; Jensen, H. B.; Poulson, R. E. "Denitrification Reactions in Shale Oil," Prepr., Div. Pet. Chem., Am. Chem. Soc., 1972, 17(4), G94.
19. Burger, E. D.; Curtin, D. J.; Myers, G. A.; Wunderlich, D. K. "Prerefining of Shale Oil," Prepr., Div. Pet. Chem., Am. Chem. Soc. 1975, 20(4), 765.
20. Burger, E. D.; Curtin, D. J.; Edison, R. R., U.S. Patent 4 003 829, January 18, 1977.
21. Curtin, D. J., U.S. Patent 4 029 571, June 14, 1977.
22. Lanning, W. C. "The Refining of Shale Oil," May 1978, U.S. Department of Energy Report BERC/IC-77/3.
23. Haunschild, W. M.; Chessmore, D. O.; Spars, B. G. "A Pilot Plant Comparison of Riser and Dense Bed Cracking of Hydrofined Feedstocks," 79th National Meeting, AIChE, Houston, Texas, March 16–20, 1975.

RECEIVED June 28, 1978.

Table XI.   Yields and Liquid Product Inspections

Whole Liquid Product N (ppm)	1325	
Run	81–4	
Feed	WOW 3394	
Run hours	556–580	
Average catalyst temperature (°F)	744	
LHSV	0.58	
Total pressure (psig)	2,202	
H$_2$ mean pressure (psia)	1,874	
Total gas in (scf/bbl)	12,304	
Recycle gas (scf/bbl)	10,308	
No loss product yields	Wt %	Vol %
C$_1$	0.42	
C$_2$	0.78	
C$_3$	0.70	
iC$_4$	0.15	0.25
nC$_4$	0.48	0.77
C$_5$–400 (°F)	10.56	13.01
400–650 (°F)	36.61	39.84
650–800 (°F)	22.95	24.53
800–end point (°F)	25.82	26.94
total C$_5$+	95.96	104.33

Appendix to Chapter 3

Tables XI and XII

Hydrotreating of Whole Shale Oil with ICR 106[a]

2800	645	505
81–4	81–4	81–4
WOW 3394	WOW 3394	WOW 3394
844–868	1252–1276	1636–1660
742	766	767
0.61	0.58	0.59
2,203	2,206	2,199
1,824	1,858	1,846
6,522	7,370	7,226
4,906	5,195	5,132

Wt %	Vol %	Wt %	Vol %	Wt %	Vol %
0.39		0.32		0.27	
0.66		0.80		0.69	
0.59		0.85		0.76	
0.13	0.21	0.19	0.31	0.17	0.28
0.38	0.61	0.57	0.91	0.53	0.85
11.19	14.29	12.36	15.23	13.24	16.32
35.08	38.89	39.32	43.78	38.72	43.08
23.18	24.64	22.61	24.27	22.56	24.24
26.29	27.19	21.76	22.93	21.72	22.91
95.75	105.03	96.06	106.22	96.26	106.57

Table XI.

Actual/no loss recovery	102.16/103.13
H_2 cons (gross) (scf/bbl)	1,996
H_2 cons (chemical) (scf/bbl)	1,923
C_5–400°F product	
gravity, API	50.6
aniline point (°F)	128.2
sulfur (ppm)	30
nitrogen (ppm)	513
low mass, LV %	
paraffins	48.7
naphthenes	38.7
aromatics	12.6
octane no., F-1 clear	
bromine no.	
viscosity, cSt, 100°F	
viscosity, cSt, 122°F	
TBP distillation (°F)	
St/5	94/206
10/30	240/295
50	336
70/90	365/395
95/99	406/428
400–650°F product	
gravity, API	33.7
aniline point (°F)	154.3
sulfur (ppm)	3.7
nitrogen (ppm)	
molecular weight	198
pour point, ASTM (°F)	5
freeze point, ASTM (°F)	
high mass, LV %	
paraffins	34.5
naphthenes	45.5
aromatics	20.0
cloud point, ASTM (°F)	
bromine no.	
viscosity, cSt, 100°F	
viscosity, cSt, 122°F	
viscosity, cSt, 210°F	
TBP distillation (°F)	
St/5	328/410
10/30	427/485
50	532
70/90	583/624
95/99	638/669

Continued

100.94/102.54	102.05/103.44	99.84/103.32
1,616	2,175	2,094
1,558	2,110	2,038
60.5	51.4	52.2
127.3	128.2	129
	14	25
506	125	67
49.4	47.3	49.4
36.8	40.7	38.9
13.8	12.0	11.6
	38.0	38.1
		0.60
	0.834	
	0.740	
85/190	99/198	60/185
215/285	222/285	213/278
330	324	322
365/401	356/392	356/395
412/463	408/467	409/444
36.8	37.5	37.4
151.7	153.7	155.5
35	14	6
388	453	356
207	202	196
5		−5
		9
34.3	33.9	33.6
45.2	46.5	46.5
20.5	19.5	19.8
	4	4
	3	2.4
3.30	3.03	
2.48	2.39	
1.25	1.18	
341/412	336/396	322/402
429/487	420/477	425/482
536	527	530
585/632	580/625	581/626
649/751	640/683	643/763

Table XI.

400°–650°F product (*continued*) ASTM D 86 distillation (°F)	
St/5	441/455
10/30	465/502
50	532
70/90	554/583
95/end point	610/616
LV % overhead	99.5
650–800°F product	
gravity, API	30.7
aniline point (°F)	184.9
sulfur (ppm)	20
nitrogen (ppm)	1,600
molecular weight	314
pour point, ASTM (°F)	
bromine no.	
viscosity, cSt, 100°F	
viscosity, cSt, 122°F	
viscosity, cSt, 210°F	
TBP distillation (°F)	
St/5	534/637
10/30	653/687
50	723
70/90	757/792
95/99	807/842
800°F–end point product	
gravity, API	26.9
aniline point (°F)	218.0
sulfur (ppm)	20
nitrogen (ppm)	1,600
molecular weight	453
pour point, ASTM (°F)	
bromine no.	
viscosity, cSt, 122°F	
viscosity, cSt, 210°F	
ASTM D 1160 distillation (°F)	
St/5	
10/30	
50	
70/90	
95/end point	
LV % overhead	
Whole liquid product	
gravity, API	34.1
aniline point (°F)	179.0
sulfur (ppm)	33
nitrogen (ppm)	1,325
molecular weight	291

Continued

446/464	431/451	
467/494	458/486	
521	514	
553/588	547/582	
597/621	597/620	
99.5	99.5	
29.9	31.5	31.6
179.3	188.0	187.3
55	13	11
3,000	960	702
318	310	310
70	75	70
	5.5	5.7
16.79	14.47	
11.17	9.97	
3.44	3.16	
407/633	463/635	479/618
652/694	651/689	642/688
731	724	727
766/801	758/793	767/
814/844	806/836	
25.5	28.4	28.6
208.7	225.8	228.6
65	10	20
4,500	995	710
472	437	452
110	115	115
	6.9	5.8
74.00	41.48	55.16
11.64	8.82	8.77
		809/821
		825/833
		858
		900/976
		1,010/1,073
		99.0
34.4	35.7	36.2
173.5	177.8	178.1
145	42	45
2,800	645	505
289	279	

Table XI.

Whole liquid product (*continued*)	
pour point, ASTM (°F)	55
bromine no.	3.9
22 component, LV %	
paraffins	38.8
naphthenes	38.2
aromatics	22.9
sulfur compounds	0.1
viscosity, cSt, 100°F	7.39
viscosity, cSt, 210°F	2.15
ASTM D 1160 distillation (°F)	
St/5	329/440
10/30	456/573
50	672
70/90	788/908
95/end point	1,016/
LV % overhead	94.0
trap, LV %	2.5
in flask, LV %	3.5

[a] Inspections of C_5–400°F do not include any C_5+ measured as gas.

Table XII.   Catalytic Cracking of

Feed nitrogen	385 ppm			
Product	Light Gasoline ST–250°F	Heavy Gasoline 250°–430°F	Light Cycle Oil 430°–625°F	Heavy Cycle Oil 650°F+
Inspection				
gravity (°API)	65.2	37.6	17.5	12.0
aniline point (°F)	113.2	< 30	< 30	
bromine no.	62	13	20	
nitrogen (ppm)	2.6	39	130	726
FIAM				
P + N	62	24		
O	31	8		
A	7	68		
Viscosity (cSt)				
at 100°F			2.45	28.18
at 130°F			1.76	13.50
Pour point (°F)			0	+70
Octane no.				
F-1 clear	89.0	95.3		
F-2 clear	79.1	84.6		

[a] Inspections of 140°–250°F fraction do not include C_5+ material measured as the gas.

Continued

80	80	85
5.9	2.8	3.0
33.8	37.8	
36.7	37.1	
29.3	24.2	
0.2	0.9	
7.42	5.730	5.57
2.17	1.860	1.88
328/422	346/428	302/397
451/569	450/561	435/550
683	667	666
791/932	769/913	765/903
968/1,021	/1,012	/1,003
95.0	93.0	93.0
4	6	6
1	1	1

Hydrotreated Shale Oils; Product Quality

870 ppm

Light Gasoline ST–250°F	*Heavy Gasoline 250°–430°F*	*Light Cycle Oil 430°–625°F*	*Heavy Cycle Oil 625°F+*
63.7[a]	39.0	20.3	18.3
110.0	< 30	< 30	
86	22	28	
9.7	141	388	1400
56	27	17	
38	13	4	
6	60	79	
		2.45	21.08
		1.83	11.50
		−20	+85
90.3	93.2		
79.5	82.8		

4

Evaluation of a Two-Stage Thermal and Catalytic Hydrocracking Process for Athabasca Bitumen

R. RANGANATHAN, B. B. PRUDEN, M. TERNAN, and J. M. DENIS

Energy Research Laboratories, Canada Centre for Mineral and Energy Technology, Department of Energy, Mines and Resources, Ottawa, Canada

A two-stage thermal and catalytic hydrocracking process of upgrading Athabasca bitumen is described. In the first stage, the bitumen is treated in the presence of excess hydrogen in a tubular reactor at 13.90 MPa (2000 psig) and 450°C. A heavy oil product is separated and a portion is recycled to the reactor. The light oil product in vapor form is further treated in a vapor-phase catalytic reactor to remove sulfur and nitrogen compounds. The second stage of this process is demonstrated by treating the light oil once-through in a bench-scale, fixed-bed catalytic reactor. The results for catalytic treatment at 13.90 MPa and 430°C suggest that the specifications of reformer naphtha feedstock, fuel oils, and cat-cracking gas-oil feedstock can be met. Three different catalysts—cobalt–molybdenum on alumina, nickel–molybdenum on alumina, and nickel–tungsten on alumina, are evaluated in the catalytic hydrocracking stage and their selectivities are compared.

The thermal hydrocracking process has been found, on the basis of pilot plant-scale runs (1,2), to give a high distillate yield and to eliminate the production of waste coke in the processing of Athabasca bitumen. This work was undertaken to further develop the thermal hydrocracking process, in keeping with the Energy Research Program of the Canada Centre for Mineral and Energy Technology, Department of Energy, Mines and Resources, and its policy of ensuring the effective use of Canada's mineral and energy resources.

0-8412-0456-X/79/33-179-053$05.00/1
Published 1979 American Chemical Society

The distillate from the thermal hydrocracking reactor contains high amounts of sulfur and nitrogen. Following the hydrocracking step, the product is distilled to produce naphtha at IBP to 204°C, light gas oil at 204°–343°C, and heavy gas oil at 343°–524°C, and these are further treated individually to remove sulfur and nitrogen compounds. Industry, by tradition, always has treated the product in individual fractions and because of fears that nitrogen compounds would migrate to the naphtha fraction, it has produced off-specification material. Once-through treatment of the combined distillates has advantages in the reduction of capital and operating costs for distillation facilities and hydrotreaters. Further reduction in costs can be obtained if the hot hydrocracked product from the first stage is treated directly without pressure let-down or condensation. It is the objective of this work to investigate the once-through hydrotreatment of combined distillate to prove the two-stage thermal and catalytic hydrocracking process (TCH process).

In the proposed TCH process, liquid feed and hydrogen would enter a thermal hydrocracking reactor and a large portion of the pitch boiling above 524°C would be converted to distillate material. The reactor products then would be treated to separate the residual pitch from the distillate. This pitch, containing most of the troublesome materials, would be withdrawn for gasification or sale as a by-product. The distillate, hydrocarbon gases, and hydrogen at high temperature and pressure would enter a second-stage catalytic reactor for desulfurization and denitrogenation.

Literature Survey

A two-stage catalytic hydrocracking process was tested by Quinsey et al. (3) for upgrading vacuum residuum from Weyburn crude. They used slurry-type catalytic hydrocracking in the first stage and a downflow fixed-bed hydrocracking in the second stage. Several reports have been published on two-stage catalytic hydrocracking of heavy oils (4,5). However, only a few authors (6,7,8) have attempted to use noncatalytic hydrocracking in the first stage. Gatsis (6) described a noncatalytic first stage where 2–30 wt % water and hydrogen were contacted countercurrently with heavy oil in a tubular reactor packed with stainless steel turnings. Schlinger et al. (7) and Brodeur and Schlinger (8) used a noncatalytic first stage where the heavy oil and hot hydrogen were contacted countercurrently. The heavy oil was fed in at the top of the contacting tower. In the above patents no mention was made of meeting the distillate feedstock and fuel oil specifications by using a two-stage upgrading process. In all cases, the heavy oil from the bottom of the first-stage contacting tower was recycled to either the top or to the mid-section of the tower.

In second-stage upgrading, the distillate oil stream is conventionally divided into naphtha at IBP to 204°C, light gas oil at 204°–343°C and heavy gas oil at 343°–524°C before hydrotreating. A few studies have been reported on treating the fractions in a single stream, and the reported results indicate that this would transfer more nitrogen compounds from high-boiling into low-boiling fractions. Frost and Poulson (9), using a downflow hydrotreating system, showed that even though the total nitrogen content of the product from one-step hydrotreating compares with that from a multistep process, the nitrogen is present in significantly higher concentrations in the naphtha and light gas-oil fractions. This poses a problem since the reformer naphtha feedstocks should contain less than 1 ppm nitrogen. In the case of sulfur removal, Frost and Poulson (9) did not observe any such difference between the single-step and multistep processes. Further, they found greater yields of naphtha and light oil fractions in the single-step process.

Experiment

The properties of Athabasca bitumen are listed in Table I. Of interest are the high sulfur, nitrogen, and ash contents compared with conventional crude. The bitumen contains 51.5% of material boiling above 524°C, 4.48 wt % sulfur, and 0.5 wt % nitrogen. The schematic of the proposed two-stage TCH process is shown in Figure 1. The bitumen and hydrogen are mixed and flow up through a tubular reactor. The temperature is maintained at 450°C and the pressure at 13.9 MPa. The mixture of gas and liquid from the top of the tubular reactor enters the

Table I. Properties of Athabasca Bitumen and Conventional Crude Oil

	Athabasca Bitumen	Conventional Crude Oil
Specific gravity, 15/15°C	1.009	0.84 to 0.9
Sulfur (wt %)	4.48	0.1 to 2.0
Ash (wt %)	0.59	nil
Conradson carbon residue (wt %)	13.3	1 to 2
Pentane insolubles (wt %)	15.5	—
Benzene insolubles (wt %)	0.72	—
Vanadium (ppm (wt))	213	2 to 10
Nickel (ppm (wt))	67	(total metals)
Total acid number	2.77	—
Total base number	1.89	—
Carbon (wt %)	83.36	86
Hydrogen (wt %)	10.52	13.5
Nitrogen (Dohrmann microcoulometer) (wt %)	0.43	0.2
Chlorine (wt %)	0.00	—
Viscosity at 38°C (cSt)	10,000	3 to 7
Pitch (524°C+) (wt %)	51.5	1 to 5

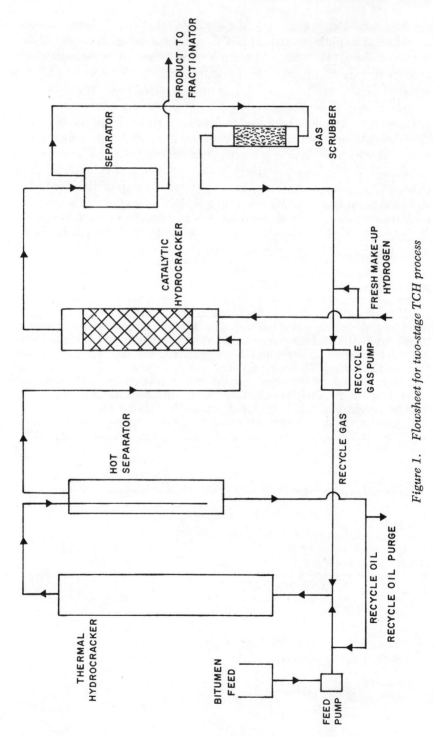

Figure 1.   Flowsheet for two-stage TCH process

hot separator which is kept at the same temperature as the reactor. Liquid from the hot separator is recycled to the tubular reactor, with a portion continuously withdrawn to prevent accumulation of metals and mineral matter. The gases from the hot separator are sent to an upflow fixed-bed catalytic hydrocracker which is the second stage of the TCH process. The pressure in the catalytic reactor is the same as in the tubular reactor but the temperature is selected to meet the product specifications. The details of typical reaction conditions in the first- and second-stage reactors are given in Table II.

In the work discussed here, the second-stage catalytic hydrocracking was simulated using a bench-scale reactor unit previously described in detail (10, 11, 12). The reactor was sequentially filled from the bottom with 42 mL of berl saddles, 100 mL of catalyst pellets, and 13 mL of berl saddles. In a typical run, feed and hydrogen flowed up through the fixed catalyst bed. Gas and liquid reaction products were separated downstream. After steady-state conditions had prevailed for 1 hr, liquid product was collected for 2 hr. The hydrogen used was 99% pure but, in an actual TCH process, recycle gas would contain some hydrogen sulfide and hydrocarbon gases. However, this would not affect the results as the recycle gas purity is reported to have no effect on the product quality (13).

Previous studies (12) have shown that catalytic pretreatment is important in obtaining stable catalytic activity. Pretreatment was carried out with feedstock at reaction conditions of 13.9 MPa (2000 psig), 400°C liquid hourly space velocity (LHSV) of 2.0 and 833 m³/t (5000 cf/bbl at 1 atm and 15°C) hydrogen for 2 hr.

The reaction studies were carried out at various temperatures. The stability of the catalyst was checked at the end of a series by repeating the first experiment in the series.

Feed for this bench-scale unit was obtained from the thermal hydrocracking experiment described previously (14). It consisted of the light ends from pilot plant run R-2–1–2 with properties as in Table IV. This

Table II.   Thermal and Catalytic Hydrocracking Conditions

Thermal Hydrocracking

Pressure (MPa)	13.9
Temperature (°C)	450
LHSV	1.0
Feed rate (g/hr)	4500
Temperature of hot separator (°C)	450
Hydrogen (m³/t)	1218
Recycle oil (g/hr)	9060

Catalytic Hydrocracking

Pressure (MPa)	13.9
Temperature (°C)	400 to 460
Catalyst loading (mL)	100
LHSV	1.0
Hydrogen (m³/t)	883

Table III.   Results from First-Stage Thermal Hydrocracking
of Athabasca Bitumen[a]

Total Liquid Product

Yield (wt %)	92.3
Yield (vol %)	99.7
Sulfur (wt %)	2.38
Nitrogen (wt %)	0.39

−524°C Distillate (wt %)

Yield	81.1
Sulfur	2.16
Nitrogen	0.25
Ash	0.00
Metals	0.00

+524°C Pitch (wt %)

Yield	11.2
Sulfur	4.5
Nitrogen	1.58
Ash	6.14
Metals	0.216

Pitch conversion (wt %)	81.5
Sulfur removal (wt %)	52.2
Nitrogen removal (wt %)	17.8
Hydrogen consumed (g mol/kg feed)	7.76

[a] All yields based on feed.

light-ends stream contained about 86% of the total distillate product resulting from fractionation in the hot separator which had only one stage of separation. The recycle-oil purge stream contained the remainder. In an operating plant this stream normally would be distilled to yield the remainder of the distillate, which would be fed to the inlet of the second-stage catalytic hydrocracker. Three commercial catalysts; i.e., 3 wt % CoO–15 wt % MoO_3 on alumina (Harshaw Chemical Co. HT-400E), 3.8 wt % NiO–15 wt % MoO_3 on alumina (American Cynamide Co. Aero HDS-3A), and 6 wt % Ni–19 wt % W on alumina (Harshaw Chemical Co. 4301E), were evaluated in the bench-scale unit. These catalysts were chosen because of their wide use in commercial hydrotreating.

It should be noted that sulfur and nitrogen balances could not be obtained between total liquid and its distilled fractions for the data presented in Tables IV, V, VI, and VII. This might be caused by the removal of disolved gases during distillation. To avoid this problem, nitrogen usually is bubbled through the samples before analysis. This could not be done for the samples obtained in this work because the naphtha might be selectively removed. However, the comparisons and conclusions should not be affected as they are based on the results of distilled fractions.

Results and Discussion

Thermal and Catalytic Hydrocracking. Typical results for the total product from the first-stage thermal hydrocracking are given in Table III. Pitch conversion of 81.5% and sulfur removal of 52.2% were obtained and only 17.8% nitrogen was removed. The feed from the first stage to the catalytic hydrocracker would have properties as in Table IV. In this table, the total feed is reported in terms of its overall properties and the properties of its constituent naphtha, light gas oil and heavy gas oil fractions. Even though some of the troublesome constituents were hydrocracked into less refractory constituents during thermal hydrocracking, the distillates still contained unacceptable amounts of sulfur and nitrogen compared with conventional crude oil distillates.

Table IV.   Properties of Oil Entering Catalytic Hydrocracking Unit

Boiling range (°C)	IBP–510
Specific gravity, 15/15°C	0.872
Sulfur (wt %)	1.97
Nitrogen (ppm (wt))	1950

Distilled Fractions

Naphtha (IBP–204°C)	
Content (vol %)	32.7
Specific gravity, 15/15°C	0.768
sulfur (wt %)	0.88
Nitrogen (ppm (wt))	575
Light Gas Oil (204–343°C)	
Content (vol %)	46.0
Specific gravity, 15/15°C	0.897
Sulfur (wt %)	2.18
Nitrogen (ppm (wt))	1404
Heavy Gas Oil (343–510°C)	
Content (vol %)	21.3
Specific gravity, 15/15°C	0.991
Sulfur (wt %)	2.82
Nitrogen (ppm (wt))	4542

Results for catalytic hydrocracking at a LHSV of 1.0, 13.9 MPa, and at several temperatures, with a 3.0 wt % CoO–12 wt % MoO₃ on aluminum catalyst, are shown in Table V. Except for the sulfur contents, the 400°C experiment showed reproducible results at the beginning and completion of the series. This shows that the catalyst was not deactivated during the experimental sequence lasting 12 hr. The volumetric and weight yields were higher than 91.3% and 86.4%, respectively. Compared with the composition of the feed to the catalytic hydrocracking unit, the naphtha and light gas oil contents increased considerably. The heavy gas oil content decreased from 21.3 vol % to as low as 5.9 vol %. The specifications for reformer naphtha feedstocks, fuel oils, and catalytic cracking gas oil are also listed in Table V. The fractions obtained from the hydrocracked product easily met all the specifications except for the sulfur content in the naphtha fraction. The

Table V.   Properties of Product from Catalytic

| | Reaction Temperature | |
Properties	*400°C*	*450°C*
Total Liquid Product		
Specific gravity, 15/15°C	0.841	0.812
Sulfur (wt %)	0.38	0.15
Nitrogen (ppm (wt))	137	11
Yield (vol %)	—	99.9
Yield (wt %)	—	93.0
Distillate Fractions Naphtha (IBP–204°C)		
Content (vol %)	33.2	49.3
Specific gravity, 15/15°C	0.761	0.762
Sulfur (wt %)	0.11	0.04
Nitrogen (ppm (wt))	7	0.5
Light Gas-Oil (204–343°C)		
Content (vol %)	53.1	44.9
Specific gravity, 15/15°C	0.869	0.862
Sulfur (wt %)	0.15	0.08
Nitrogen (ppm wt))	46	6
Heavy Gas-Oil (343°C+)		
Content (vol %)	13.7	5.9
Specific gravity, 15/15°C	0.924	0.905
Sulfur (wt %)	0.28	0.07
Nitrogen (ppm (wt %))	494	91

ᵃ From Ref. *15.*
ᵇ From Refs. *16, 17.*

nitrogen content data did not show any transfer of nitrogen compounds from higher-boiling fractions. Frost and Poulson (9) reported that, in once-through hydrocracking of shale oil on a Ni-Mo catalyst, the IBP to 177°C fraction contained about 83 ppm nitrogen compared with 5 ppm nitrogen in the 177°–288°C fraction. On the contrary, the results with Co–Mo on alumina catalyst in this work showed that naphtha contained less nitrogen than the light gas-oil fraction. The differences, discussed in the next section, could be caused by differences in the catalysts.

In the once-through studies reported in the literature, a downflow reactor scheme was used for catalytic hydrocracking (9) in contrast to an upflow reactor scheme used in this study. It has been reported in the literature that an upflow reactor scheme is superior to the usual trickle-bed operation for residual feedstocks (18, 19). Desulfurization, denitrogenation, and demetallization conversions were better in an upflow reactor.

Hydrocracking Unit (13.9 MPa and LHSV = 1.0)

Reaction Temperature		Specifications for Typical Feeds and Oils
430°C	*400°C*	
0.825	0.839	
0.15	0.17	
15	124	
91.3	97.4	
86.4	92.9	
		Reformer naphtha[a]
41.2	33.8	—
0.764	0.761	—
0.04	< 0.04	0.002
0.5	9	1
		Fuel oils[b]
52.3	52.8	—
0.867	0.869	> 0.8762
0.06	< 0.04	< 0.5
10	65	< 300
		Cat-cracking gas-oil[c]
6.5	13.4	—
0.901	0.919	—
< 0.04	0.10	< 0.5
98	—	< 2000 total N and < 100 ppm Basic N

[a] From Ref. 17.

Montagna and Shah (19) conducted experiments up to 100 hr and showed that for shallow beds the upflow reactors gave better conversions than downflow reactors and that the results were not time dependent. Specchia et al. (20) reported that their results indicated greater interfacial area and liquid mass transfer coefficients in an upflow reactor compared with downward flow. Hence, it is possible that the specifications were easily met in this work because an upflow catalytic hydrocracking reactor was used. It should be noted that at a 450°C reaction temperature and a LHSV of 1.0, the feed would be vaporized completely and the reactor would be operating as a vapor-phase reactor. At this temperature, both upflow and downflow reactors should give similar results. However, for lower temperatures, some of the feed would be in a liquid phase and hence, because of variations in liquid hold-up and back mixing, the upflow reactor system would give better results.

Comparison of Various Catalysts. Three different catalysts were evaluated at less severe hydrocracking conditions (LHSV = 2.0) to compare the relative activities and selectivities. The experiments were conducted at a LHSV of 2.0, 13.9 MPa, 883 m^3/t hydrogen and at different temperatures. The three catalysts studied were 3 wt % CoO–15 wt % MoO_3 on alumina, 3.8 wt % NiO–15 wt % MoO_3 on alumina, and 6 wt % Ni–19 wt % W on alumina. Typical results at 460°C are shown in Table VI. The results show that the three catalysts are equally active in nitrogen and sulfur removal at this temperature. However, naphtha fractions for the products obtained from Ni–Mo and Ni–W catalysts contained more nitrogen than the light gas-oil fractions. In is interesting to note that Frost and Poulson (9) also had used a Ni–Mo catalyst in the hydrocracking experiments where they observed higher amounts of nitrogen compounds in the naphtha fractions. It appears then that the phenomenon of transfer of nitrogen compounds to lower-boiling fractions depends on the type of catalyst used. This might be associated with the selectivity of Ni–Mo and Ni–W catalysts for hydrocracking. The Ni–Mo and Ni–W catalysts appeared to be more selective in hydrocracking because the yields of naphtha fraction were higher than those for the Co–Mo catalyst. Especially with the Ni–W catalyst, the naphtha fraction was extremely high. However, Table VII shows that at a 400°C reaction temperature, the difference in hydrocracking selectivity was not observed. On the other hand, at this temperature, the denitrogenation and desulfurization activities were higher for Ni–Mo and Ni–W catalysts. The results at different reaction times (Table VII) show that the denitrogenation activity dropped much faster with the Ni–W catalyst than with the other two catalysts. The nitrogen contents for the products from Ni–W and CO–Mo experiments were similar after 12 hr of reaction time. Surprisingly, the desulfurization and hydrocracking activities were

Table VI. Comparison of Different Catalysts for
Catalytic Hydrocracking[a]

	Catalysts		
Properties	3 wt % CoO–15 wt % MoO₃ On Alumina	3.8 wt % NiO–15 wt % MoO₃ On Alumina	6 wt % Ni–19 wt % W on Alumina
Total Liquid Product			
Specific gravity, 15/15°C	0.821	0.817	0.790
Sulfur (wt %)	< 0.04	< 0.04	< 0.04
Nitrogen (ppm (wt))	110	110	81
Yield (vol %)	98.5	97.2	99.7
Yield (wt %)	92.8	91.1	90.0
Naphtha (IBP–204°C)			
Content (vol %)	45.7	49.1	64.5
Specific gravity, 15/15°C	0.763	0.762	0.757
Sulfur (wt %)	< 0.04	< 0.04	< 0.04
Nitrogen (ppm (wt))	16	19	8
Light Gas Oil (204–343°C)			
Content	51.2	48.0	32.6
Specific gravity, 15/15°C	0.871	0.865	0.853
Sulfur (wt %)	< 0.04	< 0.04	< 0.04
Nitrogen (ppm (wt))	47	9	3
Heavy Gas Oil (343°C+)			
Content (vol %)	3.1	2.9	2.9
Specific gravity, 15/15°C	0.916	0.901	0.910
Sulfur (wt %)	< 0.04	< 0.04	< 0.04
Nitrogen (ppm (wt))	301	—	105

[a] Reaction conditions: 13.9 MPa, 460°C, LHSV = 2.0, 883 m³ API/t.

not significantly affected by time on stream. However, the effect on these activities would be noticeable if longer life studies were carried out. The variations in selectivities could be caused in part by different methods of preparation, because these catalysts were obtained from different manufacturers.

Several reports in the literature (15, 21, 22) also suggest differences in selectivities of Ni–Mo, Co–Mo, and Ni–W catalysts. Thomas (15) reported that Ni–Mo is superior to Co–Mo in denitrogenation. Silver et al. (21) found that Ni–W catalyst is more selective in converting quinoline-type nitrogen compounds than is the Co–Mo catalyst. However, the reported catalyst selectivities are not based on life studies. It is possible, because of poisoning, that the catalysts might not sustain the high initial activities and selectivities for a long period.

Table VII.   Comparison of Data for Various Catalysts at

Properties	3 wt % CoO–15 wt % MoO₃ on Alumina	
Reaction Time (hr)	2	12
Total Liquid Product		
Specific gravity, 15/15°C	0.845	0.841
Sulfur (wt %)	0.12	0.16
Nitrogen (ppm (wt))	369	392
Yield (vol %)	98.6	99.6
Yield (wt %)	95.6	96.1
Naphtha (IBP–204°C)		
Content (vol %)	34.6	33.9
Specific gravity, 15/15°C	0.763	0.758
Sulfur (wt %)	< 0.04	< 0.04
Nitrogen (ppm (wt))	54	38
Light Gas Oil (204°–343°C)		
Content (vol %)	48.9	52.8
Specific gravity, 15/15°C	0.876	0.873
Sulfur (wt %)	0.06	< 0.04
Nitrogen (ppm (wt))	234	238
Heavy Gas Oil (343°C+)		
Content (vol %)	16.5	13.3
Specific gravity, 15/15°C	0.932	0.937
Sulfur (wt %)	0.10	0.18
Nitrogen (ppm (wt))	1101	1308

Conclusions

Results showed that the two-stage TCH process could be used for upgrading Athabasca bitumen and for producing reformer naphtha feedstock, fuel oils, and catalytic cracking gas-oil feedstock. Product weight yields ranging from 86.4% to 93.0% were obtained. A 3 wt % CoO–15 wt % MoO₃ on alumina catalyst was found to be sufficiently active to produce specification distillates. Comparison of various catalysts showed some differences in selectivities. However, extended life studies should be carried out to substantiate the differences.

Acknowledgment

The authors wish to thank R. W. Taylor, G. R. Lett, E. Kowalchuk, R. E. Gill, M. Channing, and D. Sorenson for technical assistance.

400°C, LHSV = 2.0, and for Different Reaction Times

3.8 wt % NiO–15 wt % MoO₃ on Alumina		6 wt % Ni–19 wt % W on Alumina	
2	*12*	*2*	*12*
0.838	0.841	0.836	0.849
0.04	< 0.04	< 0.04	< 0.04
86	153	108	391
99.6	99.4	99.1	96.7
95.7	95.8	95.0	94.2
39.1	33.1	35.9	38.8
0.769	0.758	0.761	0.771
< 0.04	< 0.04	< 0.04	< 0.04
12	5	20	56
47.4	51.2	49.7	43.4
0.874	0.869	0.869	0.879
< 0.04	< 0.04	< 0.04	< 0.04
17	36	34	194
13.5	15.7	14.4	17.8
0.916	0.920	0.918	0.939
< 0.04	< 0.04	< 0.04	< 0.04
—	—	—	1083

The column headers show 3.8 wt % NiO–15 wt % MoO_3 on Alumina and 6 wt % Ni–19 wt % W on Alumina.

Literature Cited

1. Merrill, W. H.; Logie, R. B.; Denis, J. M. Mines Branch Research Report, R-281, Department of Energy, Mines and Resources, Ottawa, Canada, 1973.
2. Pruden, B. B.; Logie, R. B.; Denis, J. M.; Merrill, W. H. CANMET Report 76-33, Department of Energy, Mines and Resources, Ottawa, Canada, 1976.
3. Quinsey, D. H.; Merrill, W. H.; Herrmann, W. A. O.; Pleet, M. P. *Can. J. Chem. Eng.* 1969, 47(4), 418–421.
4. Richardson, F. W. "Oil from Coal"; Noyes Data Corporation: Park Ridge, NJ, 1975; p 268.
5. Johnson, A. R.; Alpert, S. B.; Lehman, L. M. *Proc., Div. Refin., Am. Pet. Inst.* 1968, 48, 815–840.
6. Gatsis, J. G. P.S. Patent 3 453 206, July, 1969.
7. Schlinger, W. G.; Brodeur, C. H.; Marion, C. P. U.S. Patent 3 224 959, December, 1965.
8. Brodeur, C. H.; Schlinger, W. G. U.S. Patent 3 354 076, November, 1967.
9. Frost, C. M.; Poulson, R. E. *Am. Chem. Soc. Div. Fuel. Chem., Prepr.* 1975, 20(2), 176.

10. Soutar, P. S.; McColgan, E. C.; Merrill, W. H.; Parsons, B. I. Mines Branch Technical Bulletin TB179, Department of Energy, Mines and Resources, Ottawa, Canada, 1973.
11. Williams, R. J.; Draper, R. G.; Parsons, B. I. Mines Branch Technical Bulletin TB187, Department of Energy, Mines and Resources, Ottawa, Canada, 1974.
12. Ranganathan, R.; Ternan, M.; Parsons, B. I. CANMET Report 76-15, Department of Energy, Mines and Resources, Ottawa, Canada, 1976.
13. Shah, A. M.; Pruden, B. B.; Denis, J. M. CANMET Laboratory Report ERP/ERL 77-10 (R), 1977.
14. Khulbe, C. P.; Pruden, B. B.; Denis, J. M.; Merrill, W. H. CANMET Laboratory Report ERP/ERL 76-76 (R), 1976.
15. Thomas, C. L. "Catalytic Processes and Proven Catalysts"; Academic: New York, 1970; p 158.
16. Annual Book of ASTM Standards, "Petroleum Products and Lubricants (1)," Am. Soc. Test. Mater., Annu. Book ASTM Stand. 1974, D396, D975.
17. Montgomery, D. P. Ind. Eng. Chem., Prod. Res. Dev. 1968, 7(4), 274–282.
18. Takematsu, T.; Parsons, B. I., Mines Branch Technical Bulletin TB161, Department of Energy, Mines and Resources, Ottawa, Canada, 1972.
19. Montagna, A.; Shah, Y. T. Chem. Eng. J. (Lausanne) 1975, 10, 99–105.
20. Specchia, V.; Sicardi, S.; Gianetto, A. A.I.Ch.E. J. 1974, 20(4), 646–653.
21. Silver, H. F.; Wang, N. H.; Jensen, H. B.; Poulson, R. E. Am. Chem. Soc. Div. Fuel Chem., Prepr. 1974, 19(2), 147–155.
22. Delcos, J.; Presson, R. D.; Shaw, W. G.; Strecker, H. A. Prep., Div. Pet. Chem., Am. Chem. Soc. 1966, 11(3), 185–191.

RECEIVED June 28, 1978.

Catalytic Cracking of Asphalt Ridge Bitumen

J. W. BUNGER, D. E. COGSWELL, and A. G. OBLAD

College of Mines and Mineral Industries, University of Utah,
Salt Lake City, UT 84112

*Theoretical considerations of the molecular structure of
Uinta Basin, Utah tar sand bitumen, viz. high-molecular-
weight naphthenic hydrocarbons, suggest potential for direct
catalytic cracking of these materials. Preliminary results
show that virgin bitumen is responsive to catalytic cracking
but that the efficiency of an 82 Å pore diameter catalyst is
low owing to either exclusion of molecules or adsorption of
nonvolatil molecules at the catalyst pore mouth, or both.
Typical yields at 460°–500°C and 2:1 cat-to-oil ratio are 7
wt % gas, 80 wt % liquids, and 13 wt % coke. Liquid products typically are comprised of 66% saturated hydrocarbons
and 34% aromatic and polar molecules. The gasolines produced exhibit low octane numbers resulting from competing
thermal reactions under the conditions studied. An increase
in cat-to-oil ratio results in a dramatic increase in gasoline
yield and a significant increase in coke yield at the expense
of medium and low volatility liquids. Catalytic activity
improves notably when visbroken, deasphalted, or hydropyrolyzed bitumen is used as the feedstock. For the latter
case, in which 2.6 wt % hydrogen is added during hydropyrolysis, yields are 30% gases, 64% liquids, and 6%
coke, based on the virgin bitumen.*

The bitumen contained in Utah tar sands represents a significant hydrocarbon resource. Commercial development of this resource has not
been realized principally because the highly viscous nature of the bitumen
causes major problems with recovery. Viscosities of Uinta Basin, Utah
bitumens are typically an order of magnitude greater than for the
Athabasca, Canada bitumen (1,2). The Uinta Basin bitumens, which
represent over 10 billion barrels, in place (3), are unusual also because

0-8412-0456-X/79/33-179-067$05.00/1

they are thought to be of nonmarine origin (4). The differences in origin and geochemical factors between these bitumens and those of marine origin (Tar Sand Triangle, Utah and Athabasca, Canada) has resulted in a bitumen of notably higher molecular weight, but of lower aromaticity. The predominant carbon type present in these bitumens is alicyclic, or naphthenic, carbon.

Knowledge of the structural features of Uinta Basin tar sand bitumen prompted us to take a look at the potential of these bitumens as a feed-stock for catalytic cracking. Naphthenic gas oils give high yields of high-octane gasoline (5). The high average molecular weight (approximately 700), carbon residue (about 10%), and nitrogen content (about 1%) of Uinta Basin bitumen are expected to add some difficulty to the process; however, considerations of molecular structure suggest that catalytic cracking may be an attractive alternative to thermal processing as the primary bitumen upgrading process.

In this chapter, we present the results of our exploratory evaluation of catalytic cracking of an Asphalt Ridge bitumen. The Asphalt Ridge deposit is found in the northeastern portion of the Uinta Basin and is believed to be generally representative of other Uinta Basin tar sands. Various experimental designs, dictated in part by the viscous nature of the bitumen and not necessarily representative of commercial process conditions, were used in an attempt to identify the major chemical and physical factors influencing the results. In addition to catalytic cracking of virgin bitumen, selected products derived from primary processing of bitumen were subjected to catalytic cracking to ascertain the possible role catalytic cracking might play in secondary bitumen processing. Products from catalytic cracking are compared with results obtained from thermal cracking in a bench-scale coker. The results given in this chapter are exploratory in nature and should not be construed to imply that optimum process conditions have been found.

Experimental

Bitumen and Product Feedstocks. The Asphalt Ridge tar sand was obtained from the asphalt quarry near Vernal, Utah. The bitumen was extracted in a modified soxhlet extractor and filtered through Whatman #42 filter paper. The benzene solvent was stripped from the bitumen by flash evaporation in a rotary evaporator until final conditions of 4–5 torr pressure and 90°C had been maintained for 1 hr. Previous work (6) had shown that under these conditions less than 2% bitumen was stripped as light ends and less than 0.3% benzene remained in the bitumen. The resulting extract was characterized by physical properties and elemental analysis and used directly as the charge material for the cracking reactions. Three primary products derived from Asphalt Ridge bitumen also were used as feedstocks for catalytic cracking. These are: (1) visbreaker products produced in a tubular reactor at conditions of 425°C, 150 psig, and 11 min space time; (2) maltenes prepared by dissolution of bitumen in 40 volumes of *n*-pentane under magnetic stirring for 1 hr at 30°C, followed by overnight digestion at 20°C and filtration of the mixture using minimal amounts of chilled (about 10°C) pentane for washing; and (3) hydropyrolysis liquid products produced in a tubular reactor at 525°C, 1500 psig H_2, 18 sec residence time, and 2 LHSV according to procedures described more fully by Ramakrishnan (7).

Batch-Type Catalytic Cracking. Batch-type reactions were carried out at atmospheric pressure in a 316 stainless steel reactor resembling a flash distillation apparatus with sidearm. The reaction chamber was fitted with a Vycor glass liner. Approximately 10 g of bitumen, diluted with benzene, and a prescribed amount of equilibrium catalyst was added to the glass liner and the solvent subsequently was stripped at conditions described above. Such a design is thought to provide ultimate contact of bitumen with catalyst. The reactor was immersed in a preheated fluidized-bed sand bath. The reaction was allowed to go to completion (about one-half hour) at which time coke, liquid condensate, and gas yields were determined gravimetrically. Gas yields were measured directly by initially collecting the product gases in a water trap and by immediately determining the density of the gases on a Mettler DMA-40 densitometer after the run. The weight of gas produced then was calculated from the density and volume of the gas at atmospheric pressure (about 640 mm Hg) and room temperature (approximately 20°C).

Semibatch Catalytic Cracking. These reactions were designed to approximate conditions used in the Cat-A test (8). The reaction zone of the vertically mounted all-glass reactor consisted of a 1 in. × 6 in. Vycor glass tube that was surrounded by a copper sheath and heated in a Lindberg furnace. The bitumen was contacted in a downflow mode and yields were determined gravimetrically, as described above. Because a small amount of unreacted bitumen or partially pyrolyzed residue remained in the inlet tubes after the reaction period and were counted as coke by this procedure, an additional attempt was made to determine the actual coke yield on the catalyst. The coke yield was determined by measuring the weight loss when coke was burned off the catalyst at 600°C. This procedure tends to underestimate the coke yield because the partially pyrolyzed residue is not included.

Product Evaluation. Gaseous products were analyzed by gas chromatography (GC) using an alumina column. Liquid products were evaluated by physical properties, elemental analysis, liquid chromatographic separation (9), and simulated distillation (10). Distillation to produce a gasoline fraction was accomplished with a spinning band distillation column. Gasoline analysis and evaluation was obtained from an industrial analysis laboratory.

Results and Discussion

Characterization of Feedstock and Catalysts. The elemental analysis and physical properties of various feedstocks used in this study are given in Table I. The data show that the virgin bitumen contains

Table I.   Elemental Analysis and Properties of Catalytic
Cracking Feedstocks (Asphalt Ridge Bitumen)

	Virgin Bitumen	Vis-breaking Products	Maltenes	Hydro-pyrolysis Products
Carbon (wt %)	86.3	86.0	86.9	87.1
Hydrogen	11.3	11.3	11.3	11.5
Nitrogen	1.1	1.0	0.8	0.8
Sulfur	0.4	0.2	0.4	0.3
Oxygen	0.9	1.5	0.6	0.3
H/C (atomic)	1.56	1.58	1.55	1.57
Average molecular weight (VPO—benzene)	713	477	600	321
API gravity	12.5	14.7	14.2	25.2
Carbon residue (Conradson)	9.1	N.A.	6.5	1.4
% Volatile at 538°C nominal boiling point	40	67	45	85

appreciable quantities of nitrogen, only moderate amounts of sulfur, and is of high average molecular weight and low volatility. Comparison of the data with that derived for other Uinta Basin bitumens (1, 10, 11) shows that none of the Asphalt Ridge bitumen properties is atypical. Elemental composition of the visbroken liquids is virtually unchanged from the virgin bitumen; however, the average molecular weight has decreased, and a corresponding increase in volatility is realized. The maltene fraction, which represents 90% of the original bitumen, has a notably lower content of heteroatoms but exhibits no essential difference in H/C ratio when compared with the virgin bitumen. The maltene fraction also exhibits a lower average molecular weight and carbon residue and a higher API gravity and volatility compared with the total bitumen. The hydropyrolysis products, which represent 73% of the bitumen (27% hydrocarbon gases are produced in this process at the conditions used (12)), exhibit elemental analysis similar to the maltenes but a lower molecular weight and carbon residue and a higher API gravity and volatility than the maltenes.

Previous analysis of Uinta Basin bitumens (6, 10) have shown that the predominant nitrogen types are pyrollic, amide, and aromatic nitrogen. Predominant sulfur types are sulfide, sulfoxide, and thiophenic sulfur; predominant oxygen types are ketones, phenols, carboxylic acids, and possibly appreciable concentrations of furans. The aromatic (basic) nitrogen is expected to participate in irreversible adsorption on acid-cracking catalysts (13) and it could require an increased cat-to-oil ratio. Previous analyses (10) have indicated also that Uinta Basin bitumen is high in naphthenic hydrocarbon and low in free paraffins. This is illustrated by the results given in Table II in which over 60% of the saturates

Table II. Group-Type Analysis of P. R. Spring
Saturated Hydrocarbons (10)[a]

Number of Rings	Wt % of Saturates
0	7.1
1	12.3
2	29.4
3	31.5
4	14.1
5	4.4
6	1.3
Monoaromatics	0

[a] 26 wt % of total bitumen, VPO—av mol wt = 325.

(16% of the bitumen) consists of perhydronaphthalenes, perhydrophenanthrenes, perhydroanthracenes, and perhydrofluorenes. The generally naphthenic character is thought to persist in the aromatic portions of the total bitumen also.

The measured molecular weight of 713 for the virgin bitumen is considered to be higher than the true molecular weight (1, 6) because of the indeterminate effects of solute–solute intermolecular association. Accurate measurements of molecular weight by mass spectrometry (MS) have not been achieved with tar sand bitumens because of physical limitations resulting from the low volatility of these materials. Considering, however, that the initial boiling point for Asphalt Ridge bitumen is about 200°C and that only 40% exhibits a TBP below 538°C it is unlikely that the true average molecular weight is less than about 450. Thus, tar sand bitumen is comprised of significantly larger molecules on the average than conventional cat cracker feedstocks, atmospheric and vacuum gas oils, which are 100% distillable below the nominal boiling point of 538°C. From previous experience with liquid chromatographic separations of bitumen samples (14), it was felt that a catalyst possessing a pore diameter of less than about 40 Å effectively would exclude bitumen molecules from the interior of the catalyst particles. It was principally on the basis of this pore size consideration that the catalysts were selected for this study.

Two catalysts were selected for application—an amorphous silica–alumina catalyst possessing an 82-Å average pore diameter and a molecular sieve catalyst possessing a 200-Å average pore diameter for the superstructure. These catalysts are thought to be representative of commercial cracking catalysts capable of handling high-molecular-weight feeds. Specifications for these catalysts are given in Table III. Catalysts were used as equilibrium catalysts, that is, catalysts which have been recovered from commercial cracking units. For certain experiments the catalyst was ground and sized to include about equal quantities of 25–35,

Table III. Properties of Catalysts

Composition	(15) Molecular Sieve (Mobil Durabead-8)	(16) Silica–Alumina (Houdry 159CP)
Al_2O_3	48.4	12.4
SiO_2	49.5	87.3
Na_2O	0.6	0.6
Physical Properties		
surface area (m^2/g) (BET)	70	290–315
absorption (wt %)	—	58
porosity (vol %)	33	57
average pore diameter (Å)	200	82
particle size (mm)	about 4 mm sphere	4 × 4
bulk density (g/mL)	0.96	0.62
real density (g/mL)	—	2.2–2.4
pellet density (g/mL)	—	0.99

35–200, −200 mesh catalyst. The purpose of using finely ground catalysts was to qualitatively assess possible effects of diffusion and to assess whether or not adsorption or coke deposition on the external surface of the catalyst particle was inhibiting the activity of the interior of the catalyst particle.

The virgin bitumen is vaporized only partially at normal cracking temperatures and obtaining reproducible catalyst–bitumen contact presents a significant problem. Two modes of operation were used—semibatch and batch. In the semibatch mode, the catalyst and feed were preheated separately before contact was made in a downflow fixed-bed reactor. Preheat of the feed was held to 350°C to minimize thermal-cracking reactions prior to contact. This reactor had contact between bitumen and catalyst at reaction temperature, but it did not achieve uniform contact and meaningful cat-to-oil ratios. The batch mode provided ultimate contact with the bitumen but it had longer heat-up times

Table IV. Results of Catalytic Cracking of

Designation	Catalyst	T(°C)	Mode	Cat–Oil
Bt(2)	MS(f)	412	SB	1.8
Bt(8)	MS(f)	414	B	1.0
Bt(1)	SA	412	SB	1.3
Bt(9)	S/A(f)	412	B	1.0
Bt(10)	T	415	B	—

[a] Symbol designation: Bt (virgin bitumen); S/A (silica–alumina catalyst); MS (molecular sieve catalyst); SB (semibatch Cat A mode); B (batch mode); f (powdered catalyst); T (thermal coking).

(approximately 10 min). A more elaborate reactor design such as a continuous riser reactor was not warranted at the preliminary stage of this research.

Gravimetric Results of Catalytic Cracking. Experiments were conducted to assess the effects of temperature, cat-to-oil ratio, and feedstock composition. In addition to the effect of variables on product yields, it was also important to identify the relative influence of thermal reactions, since free-radical reactions may adversely affect product quality. A series of experiments was conducted in the temperature range of 412°–415°C because this is the temperature of maximum increase in production from thermal cracking and catalytic vs. thermal effects are more easily discernible at this temperature.

Results of the low-temperature cracking are given in Table IV along with the results obtained at 415°C from coking. Results of Bt(2) and Bt(8) show good selectivity of the molecular sieve catalyst toward production of liquids. Results of these two runs illustrate general agreement between the two reactor configurations. Comparison of results for catalytic cracking [Bt(8)] with thermal cracking [Bt(10)] shows that the bitumen is quite responsive to catalytic cracking in that catalytic cracking is more efficient in converting bitumen to liquid and gaseous products at 415°C.

Results of mild-temperature catalytic cracking on an amorphous silica–alumina catalyst are given in runs Bt(1) and Bt(9) and show a poorer selectivity toward production of liquids compared with results obtained with the molecular sieve catalyst. Conversion of high-molecular-weight species to liquid products is notably poorer when the fine-particle-sized catalyst is used. The differences between results in Bt(1) and Bt(9) are not thought to be caused by the differences in reactor configuration because semiquantitative data obtained from using the ground catalyst in the semibatch mode also showed extremely high residue/coke yields. Comparison with results for thermal cracking [Bt(10)] show enhanced

Asphalt Ridge Bitumen at Low Temperature[a]

Weight Percent Yields			
Gas	Liquid	Residue[b] (Coke)	API Liquids
1	79	20 (11)	29.5
2	76	22	29.3
6	67	27 (14)	27.9
7	50	43	31.7
2	58	40	26.9

[b] Values in parenthesis were determined by burning the coke from the catalyst.

cracking activity when the pelleted form of the catalyst is used, but reduced activity when the ground catalyst is used. These results strongly suggest that the silica–alumina catalyst is susceptible to pore mouth and/ or adsorption of virgin bitumen on the exterior of the catalyst particles which renders the catalyst sites unavailable. The higher yields exhibited in Bt(1) compared with Bt(10) result partially from mild thermal cracking occurring in the void spaces between catalyst pellets which reduce the molecular weight prior to the catalytic cracking and from the enhanced cracking activity caused by the presence of the catalyst. Thus, it is concluded from the experiments at 412°–415°C that while the bitumen is responsive to catalytic cracking with virgin bitumen molecules, high diffusional resistances are experienced with an 82-Å pore diameter catalyst. This resistance may be intrinsic or as a result of adsorption of bitumen molecules at the surfaces effectively inhibiting further diffusion. A study of the factors influencing poor accessibility to the catalyst is important to possible utilization of catalytic cracking as a process for upgrading virgin bitumen.

A series of experiments were conducted at about 460°C. This temperature is closer to the temperature used in commercial units operating at low severity. Testing at moderate conversions minimizes the thermal effects and tends to maximize the differences experienced by a change in variables other than temperature. Results are given in Table V and show that catalytic cracking (Runs Bt(3, 4, 5)) produces yields similar to or better than coking (Bt(11)). A single run at 500°C (Bt(16)) resulted

Table V. Results of Catalytic Cracking of

Designation	Catalyst	Mode	T(°C)	Cat–Oil
Bt(3)	S/A	SB	470	1.3
Bt(4)	S/A	B	460	2.0
Bt(12)	S/A	B	460	5.0
Bt(13)	S/A	B	460	10.0
Bt(16)	S/A	B	500	2.0
M(5)	S/A	B	460	2.0
HP(14)	S/A	B	426	2.3
HP(15)	S/A	B	460	2.0
Bt(6)	MS	B	460	3.0
VB(7)	MS	B	460	3.0
Bt(11)	T	B	460	—
Bt(17)	T	B	625	—

[a] Symbol designation: M (maltenes from virgin bitumen); VB (visbroken bitumen); HP (hydropyrolyzed bitumen). See also Table IV.

in a higher production of gases plus liquids and a higher API gravity liquid than was experienced with destructive distillation terminating at 625°C [Bt(17)] (*see* Ref. *17*).

The effects of cat-to-oil ratio are shown by Runs Bt(4), Bt(12), and Bt(13). Gas yields are variable and probably reflect the effects of competing reactions; thermal reactions are almost certainly relatively more influential at low catalyst-to-oil ratios than at high catalyst-to-oil ratios because of the greater opportunity for catalytic reactions to occur at higher ratios. Total liquid and gas yields decrease with increasing catalyst-to-oil ratio but API gravity of liquids increases significantly with increasing ratio. These results are consistent with published literature on the effect of catalyst-to-oil ratios on gas oil cracking which show that total yields of gases and liquids decrease, conversion to gasoline increases, and the octane number of gasoline increases as the catalyst-to-oil ratio increases (*5*).

The effect of asphaltenes on the results are seen in the comparison of Bt(4) with M(5). The feedstock for M(5) was the *n*-pentane-deasphalted bitumen (maltenes) and represented 90% of the total bitumen. Only 4% greater yields were experienced in M(5). These results are explained by observing that the asphaltenes from tar sand bitumen consist of the high-molecular-weight species but that they also possess appreciable amounts of hydrogen ($H/C = 1.20$). The molecules which comprise the asphaltenes, largely high molecular weight but not totally aromatic molecules, actually contribute significantly to the liquid product, albeit

Asphalt Ridge Bitumen at Moderate Temperatures[a]

Weight Percent Yields

Gas	Liquid	Residue (Coke)	ÅPI Liquids
11	76	13 (6)	25.1
10	74	16	30.8
7	73	20	35.6
9	67	24	41.5
10	79	11	30.3
10	78	12	32.0
2	80	18	33.9
4	88	8	32.2
7	80	13	27.1
4	83	13	28.8
4	81	15	27.1
5	82	13	25.8

the heavy products as seen in the API gravity results. Conversely, not all coke precursors are removed by solvent deasphalting as shown by the 12% coke yield for Run M(5). Significant quantities of aromatic-containing molecules are soluble in the precipitating solvent n-pentane. (See, e.g., a comparison of compound type analysis for whole and deasphalted bitumen (1)). These results illustrate that a one-to-one precursor–product mechanism does not exist between molecules comprising the asphaltene fraction and those responsible for coke formation in catalytic cracking of tar sand bitumen. Correlations between asphaltene content and the propensity for coke formation are largely fortuitous.

Run HP(14) used a more drastically upgraded bitumen as a feedstock. The feedstock, in this case, was produced by hydropyrolysis in a tubular reactor. The feedstock possessed an elemental composition quite similar to the original bitumen; however, the average molecular weight of the feed was 321, or less than half that of the original bitumen. Results of HP(14) show excellent responsiveness and selectivity at very mild severity. At 460°C [HP(15)] the hydropyrolysis liquids produced a total liquid plus gas yield of 92%. Considering the yields obtained in hydropyrolysis, a two-stage process of hydropyrolysis plus catalytic cracking would yield a total of over 94% gases plus distillable liquids based on the virgin bitumen, without the use of hydrogenation catalysts. The thermal addition of about 2.6 wt % H_2 during hydropyrolysis significantly reduces the amount of coke produced.

Visbreaking is a relatively inexpensive process which is potentially applicable to primary processing of tar sand bitumen. Visbreaking has the principal effect of reducing the molecular weight without greatly altering the gross structure of the bitumen. Results of catalytic cracking of a virgin bitumen on a molecular sieve catalyst at 460°C [Bt(6)] had shown a somewhat high gas yield and suggested that appreciable thermal reactions were occurring. When a 477-av mol wt visbroken bitumen was charged to the molecular sieve catalyst under identical conditions [VB-(7)] the result was a greater catalytic effect. This effect is apparent in the selectivity toward production of liquid products and no notable degradation of liquid product quality when the API gravity of VB(7) liquid is compared with Bt(6) liquid.

Characterization of Products. The gases and liquids from selected runs were subjected to detailed compositional analysis. Table VI gives the C_1 to C_4 gas analysis as a function of catalyst type and compares the results from cat cracking with thermal cracking. With Run Bt(6) only moderate catalytic activity was experienced as evidenced by the slightly improved C_3 and C_4 content compared with Bt(11). Thermal gases are characterized by high C_1 and C_2 yields. Somewhat better catalytic activity was experienced with Bt(4) as evidenced by the substantial

Table VI. C_1 to C_4 Gas Analysis

Run Designation[a]

	Bt(11) Thermal (460°C) (wt %)	Bt(6) MS 460°C (wt %)	Bt(4) S/A 460°C (wt %)
Methane	41.0	32.3	18.9
Ethane	16.5	16.1	10.3
Ethylene	15.1	5.0	4.2
Propane	10.0	15.4	12.2
Propylene	11.6	10.2	15.4
n-Butane	1.3	5.4	3.3
Isobutane	1.4	4.0	18.0
Butylenes	3.1	11.6	17.7

[a] *See* Tables IV and V for explanation of run designations.

percentages of C_4's produced and the high isobutane/n-butane ratio. Results of Bt(6) are almost certainly less than the optimum that can be achieved by this catalyst but illustrate the difficulty which may be encountered owing to competing reactions when cracking these high-molecular-weight bitumens.

Bulk property analysis for liquid products of selected runs are given in Table VII. Product inspections for the coker distillate [Bt(11)] are given for reference. Hydrogen-to-carbon ratios do not vary appreciably but the products from catalytic cracking possess slightly less hydrogen than the thermal products. Nitrogen and sulfur contents of products are variable and, for the virgin bitumen feed, increase with increasing temperature. Nitrogen content increases with increasing cat-to-oil ratio while sulfur content decreases as a function of this variable. The appreciable increase in nitrogen content with increased cat-to-oil ratio is not readily explained; it is not known whether the nitrogen in Bt(4) is distributed more heavily in the gases or the coke compared with Bt(13). The liquid products from the hydropyrolysis product feed show very low hetero-atom contents and the lowest average molecular weight of the liquids produced.

Liquid products were analyzed by a simulated distillation to ascertain the boiling point distribution. Results of this analysis are given in Table VIII and reveal a wide variation in boiling point distributions. Products from the low-temperature run are significantly higher boiling than products derived at higher temperatures. Between 460°C [Bt(4)] and 500°C [Bt(16)] temperature has only a minor effect on boiling point distribution. Cat-to-oil ratio has a major effect on the product distribution

Table VII. Elemental Analysis and Physical Properties

	Bt(11) T/460	Bt(4) CC/460/2:1
Carbon	86.7	87.0
Hydrogen	12.2	11.5
Nitrogen	0.48	0.24
Sulfur	0.29	0.25
Oxygen	0.31	0.79
H/C Atomic Ratio	1.68	1.58
API Gravity	27.1	29.6
Specific gravity, 20/20	0.892	0.878
Refractive index, 20°C	1.4998	1.5019
Average molecular weight	321	272

[a] See Tables IV and V for a more complete description of reaction conditions.

Table VIII. Simulated Distillation of

Equivalent BMCOA Fraction Number[a]	Cut Point °C	Cumulative Weight Percent Bt(1) 412/1.3:1	Bt(4) 460/2:1
1	50		0.9
2	75		2.4
3	100	1.2	8.3
4	125	2.3	11.2
5	150	4.2	14.9
6	175	6.3	18.9
7	200	8.7	22.8
8	225	11.0	27.3
9	250	14.7	32.5
10	275	19.4	38.2
11	305	25.2	46.0
12	335	32.5	54.9
13	365	40.3	62.9
14	395	49.5	71.0
15	425	58.8	78.6
16	455	68.8	85.2
17	485	80.4	92.1
18	515	90.6	96.6
19	538	96.6	99.1
Residue	> 538	100.0	100.0

[a] See Ref. 1 for explanation.

of Liquid Products from Thermal and Catalytic Cracking

Run Designation[a]

Bt(13) CC/460/10:1	Bt(16) CC/500/2:1	HP(15) CC/460/2:1	VB(7) CC/460/3:1
87.2	86.8	86.8	86.9
11.4	11.6	11.9	11.5
0.45	0.52	0.22	0.31
0.11	0.35	0.12	0.27
0.81	0.73	0.68	0.75
1.56	1.59	1.63	1.58
41.5	30.3	32.3	28.8
0.818	0.875	0.864	0.882
1.4831	1.5056	1.4914	1.5002
N.A.	410	245	N.A.

Liquid Products   from Catalytic Cracking

Cumulative Weight Percent

Bt(13) 460/10:1	Bt(16) 500/2:1	HP(15) 460/2:1	VB(7) 460/3:1
2.1	2.1	0.5	0.4
6.8	4.7	1.8	1.5
18.5	8.5	4.2	3.1
25.0	12.0	7.2	5.2
32.1	16.0	11.1	7.8
39.9	19.7	15.1	10.6
47.8	23.7	19.6	13.9
55.6	28.0	24.0	17.6
64.3	33.5	30.5	23.0
72.5	39.5	37.8	29.2
81.2	46.7	46.6	37.6
88.1	54.4	56.8	47.2
93.5	61.6	67.5	57.5
96.9	68.9	77.6	67.7
98.5	75.0	85.9	76.4
99.2	81.5	92.7	85.3
99.6	90.2	98.0	93.7
99.8	94.9	99.7	98.2
99.9	98.0	100.0	100.0
100.0	100.0	100.0	100.0

and, as can be seen in run Bt(13), nearly 50% of the liquids are C_5–200°C gasoline. The boiling point distribution resulting from the hydropyrolyzed bitumen feedstock is surprisingly similar to that obtained from virgin bitumen considering the differences in average molecular weight of these two feedstocks. The slightly higher boiling character of the products from the visbroken feed may be attributable to the difference in catalysts.

The liquid products from Run Bt(4) were selected for study of detailed compositional analysis. Gas analysis (Table VI) had shown that good catalytic activity had been achieved on the virgin bitumen at a moderate temperature and a low cat-to-oil ratio with this run. The products from this run were separated according to hydrocarbon classes by dual silica–alumina column chromatography. Results of this separation are given in Table IX and are compared with similar results obtained on the coker distillate. Values given in Table IX show that approximately half of the charge material was recovered as saturated hydrocarbons. Losses, amounting to over 10% of the charge, are attributable largely to volatilization of light ends, principally saturated hydrocarbons, during removal of the chromatographic solvent (see further discussion below). Only minor differences are seen in the hydrocarbon type analysis when catalytically derived products are compared with thermally derived products.

Table IX.   Compound Type Classification of Products

	Bt(4) (wt %)	Bt(11) (wt %)
Saturates	52.9	49.4
Monoaromatics	7.8	10.5
Diaromatics	4.7	7.5
Polypolar Aromatics	20.0	22.2
Loss	14.6	10.4
	100	100

The hydrocarbon type fractions for the catalytically cracked products were analyzed by simulated distillation and results are given in Figure 1. Values for the ordinate for the fraction were determined by multiplying the simulated distillation data by the weight fraction material recovered from the chromatographic separation. Thus, when the values for the fractions are totalled, the resulting curve (sum total) includes the cumulative errors of both the simulated distillation and liquid chromatographic separation. With no errors in analysis and complete recovery in the separation, the sum total curve would coincide with the original total curve.

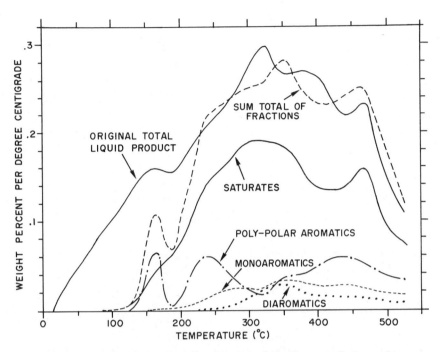

Figure 1. Boiling point distribution of products from catalytic cracking

The results in Figure 1 show that above 225°C reasonably good agreement is achieved between the original material and the sum of the fractions. Principal causes for underaccountability lie with losses during separation, both by volatilization of the sample and irreversible adsorption of highly polar species. Results indicate that almost all of the 14.6% loss is attributable to loss of light ends and that virtually all of this is saturated hydrocarbons. Polypolar aromatics, the only other major fraction, is not expected to contain large amounts of highly volatile material. Causes for overaccountability probably rest with inaccuracies with response factors in the simulated distillation procedure. Minor alteration, such as by air oxidation or polymerization of the compounds during handling and chromatographic separation, cannot be discounted as possible causes for disagreement in the results, but precautions were taken to minimize the exposure to heat, light, and air.

The results in Figure 1 can be interpreted in terms of general ring structures with the hydrocarbon classes. The peak for the polypolar aromatic fraction at 160°C probably is caused by polar–monocyclic compounds and the peak at 240°C is probably the result of polar dicyclic compounds. The broad curve in the monoaromatics centering at 275°C is probably mainly caused by alkyl-substituted tetralins while the peaks

Table X. Analysis of Gasoline from Run Bt(4) [a]

	Wt %	Vol %	RO [b]	MO [b]
n-Paraffins	5.0	4.3	−4.9	−4.5
Isoparaffins	20.7	18.2	45.2	48.5
Naphthenes	10.7	10.3	79.1	74.4
Olefins	11.1	9.8	86.4	73.6
Aromatics	40.1	45.1	99.4	89.6
Unclassified	12.4	12.3	0	0
C-5–200°F	16.8		85	74
200–392°F	83.2		70	60
Calculated blended values	100		73	62

[a] See Tables IV and V for a description of the run.
[b] RO and MO are research octane number and motor octane number, respectively.

centering at 350°C are the result of tricyclic compounds; in the case of polypolar aromatics these tricyclics may be phenanthrenes and anthracenes. Material boiling at about 440°C probably is caused by tetracyclic species. The rather intense band in the saturates at about 460°–470°C may be the result of tetra- and pentacyclic biological markers reported to be present in Asphalt Ridge bitumen by Thomas (18).

The C_5–200°C gasoline from run Bt(4) was isolated by distillation and analyzed for its composition and octane rating. Results are summarized in Table X and show somewhat low octane ratings. The low octane ratings are attributable to the presence of thermal products, n-paraffins, and only slightly branched isoparaffins. Over 12% of the product was unidentified and was assigned an octane number of zero. Results of the gasoline analysis reveal that optimum conditions were not achieved under conditions of Run Bt(4) and that further efforts must be made to produce a high-quality gasoline directly from virgin bitumen. Certainly, higher temperatures and higher cat-to-oil ratios will help; however, with high cat-to-oil ratios process costs will increase and, as the data shows, yields will decrease significantly. Attempts also should be made to contact the feed and the catalyst at preheated conditions to more closely simulate conditions present in a modern fluid catalytic cracker (FCC). The selection of catalysts used in this study was quite arbitrary and better catalysts for primary conversion of bitumen may well exist.

Conclusions

Bitumen derived from the Asphalt Ridge deposit in Utah was responsive to catalytic cracking, which provided higher quality products at similar yields when compared with coking. Good reactivity to catalytic cracking is predicted from consideration of the high content of alkyl and naphthenic carbon present. The extremely high molecular weight of virgin bitumen results, however, in an inhibited rate of catalytic cracking and allows competing thermal reactions to adversely influence the product quality under certain conditions studied. Octane numbers for the gasoline produced are low as a result of this effect. Coke make is substantially higher than is experienced presently for commercial gas–oil cracking. Higher aromatics responsible for producing coke are not removed quantitatively by prior deasphalting and deasphalting removes material which contributes to the liquid and gas yields. Mildly upgraded products with a reduced molecular weight are notably more reactive to catalytic cracking than the original bitumen. High total yields of gases, gasoline, and middle distillates are expected from hydropyrolysis followed by catalytic cracking under proper reaction conditions. Further study of more optimum reactor configurations and process conditions is indicated. Whether the feed is a virgin bitumen or, more likely, an upgraded bitumen product, catalytic cracking probably will play an important role in the commercial development of Uinta Basin tar sands.

Acknowledgments

The authors gratefully acknowledge R. Beishline and the students of Weber State College for the liquid chromatographic separation and distillation to produce the gasoline fraction. The authors wish to acknowledge the technical assistance of M. S. Shaikh during the early stages of this work. This work was supported by the U.S. Department of Energy and the State of Utah.

Literature Cited

1. Bunger, J. W. "Techniques of Analysis of Tar Sand Bitumens," *Prepr., Div. Pet. Chem., Am. Chem. Soc.* 1977, 22(2), 716–726.
2. Oblad, A. G.; Seader, J. D.; Miller, J. D.; Bunger, J. W. "Oil Shale and Tar Sands"; Smith, J. W., Atwood, M. T., Eds.; *AIChE Symp. Ser.* 1976, 72(155), 69–78.
3. Ritzma, H. R., compiler, "Oil Impregnated Rock Deposits of Utah," Utah Geological and Mineralogical Survey, Map-47, Salt Lake City, Utah, January 1979, 2 sheets.
4. Ritzma, Howard, personal communication, 1976.
5. Oblad, A. G.; Milliken, T. H.; Mills, G. A. "The Effects of Variables in Catalytic Cracking," In "The Chemistry of Petroleum Hydrocarbons"; Brooks, B. T., et al., Eds.; Rheinhold: New York, 1954; Chapter 28.
6. Bunger, J. W.; Thomas, K. P.; Dorrence, S. M. "Compound Types and Properties of Utah and Athabasca Tar Sand Bitumens," *Fuel*, 1979, 58(3), 183.
7. Ramakrishnan, R. "Hydropyrolysis of Coal Derived Liquids and Related Model Compounds," Ph.D. Dissertation, Dept. of Fuels Engineering, University of Utah, 1978.
8. Alexander, J.; Shimp, H. G. "Laboratory Method for Determining the Activity of Cracking Catalysts," *Nat. Pet. News* 1944, 36, R537.

9. Hirsch, D. E.; Hopkins, R. L.; Coleman, H. F.; Cotton, F. O.; Thompson, C. J. "Separation of High-Boiling Distillates Using Gradient Elution through Dual-Packed (Silica Gel-Alumina Gel) Adsorption Columns," *Anal. Chem.* **1972,** *44* (6), 915–919.
10. Bunger, J. W. In "Shale Oil, Tar Sands and Related Materials," *Adv. Chem. Ser.* **1976,** *151,* 121–136.
11. Wood, R. E.; Ritzma, R. R. "Analysis of Oil Extracted from Oil-Impregnated Sandstone Deposits in Utah," *Special Studies 39,* Utah Geological and Mineralogical Survey, Salt Lake City, 1972.
12. Bunger, J. W.; Cogswell, D. E.; Oblad, A. G. "Influence of Chemical Factors on Primary Processing of Utah Tar Sand Bitumen," *Am. Chem. Soc., Div. Fuel Chem., Prepr.* **1978,** *23* (4), 98–109.
13. Oblad, A. G.; Milliken, T. H., Jr.; Mills, G. A. "Chemical Characteristics and Structure of Cracking Catalysts," *Adv. Catal.* **1951,** *2,* 199–247.
14. Bunger, J. W., unpublished results obtained while the author was employed at the Laramie Energy Technology Center, Laramie, WY, 1973.
15. Eastwood, S. C.; Plank, C. J.; Weisz, P. B. "New Developments in Catalytic Cracking," *Proc., World Pet. Cong., 8th* **1971,** *4,* 245–254.
16. Rim, J. J. "Chemical Characterization of Molecular Sieve Catalysts," Ph.D. Thesis, Dept. of Fuels Eng., Univ. of Utah, 1974.
17. Bunger, J. W.; Mori, S.; Oblad, A. G. "Processing of Utah Tar Sand Bitumens, Part I. Thermal Cracking of Utah and Athabasca Tar Sand Bitumen," *Am. Chem. Soc., Div. Fuel Chem., Prepr.* **1976,** *21* (6), 147–158.
18. Thomas, K. P.; Barbour, R. V.; Guffey, F. D.; Dorrence, S. M. "Analysis of High-Molecular-Weight Hydrocarbons in a Utah Tar Sand and Produced Oils," In "The Oil Sands of Canada–Venezuela—1977," *CIM Spec. Vol. 17* **1977,** 168–174.

RECEIVED June 28, 1978.

6

Development of a Process for the Conversion of Coal to Catalytic-Cracking Charge Stock

A. SCHNEIDER, E. J. HOLLSTEIN, and E. J. JANOSKI

Suntech, Inc., P.O. Box 1135, Marcus Hook, PA 19061

Batch experiments on the hydroliquefaction of Illinois No. 6 coal in a solvent of hydrogenated anthracene oil using slurried cobalt molybdate catalysts have been carried out. Subsequent treatments of the total coal liquid products as well as the 1000°F— distillate fractions thereof with hydrogen chloride produced raffinates for catalytic cracking containing less than 0.1% nitrogen. The effects of operating conditions during catalytic hydroliquefaction on the yields of HCl–precipitatable nitrogen compounds in the total liquid products and in the 1000°F— distillate fractions have been determined. In the hydroliquefaction step, reduction of catalyst particle size increases the extent of hydrodenitrogenation and thereby, in the subsequent HCl treating, decreases the yields of amine hydrochloride precipitates.

This chapter is concerned with the development of a process for converting coal to gas oil of suitable quality for catalytic cracking. The primary emphasis is in the development of a practical method for the selective removal of nitrogen from coal liquids to give a gas-oil product containing 0.1% or less of nitrogen. This method consists of supplementing the partial hydrodenitrogenation achievable during catalytic hydroliquefaction of coal in the presence of a solvent (1) by a subsequent treatment of the filtered product with dry hydrogen chloride (2). Since the extent of hydrodenitrogenation, conversion of nitrogen to basis nitrogen, and conversion of coal to liquid in the gas oil boiling range depend on the conditions during the catalytic hydroliquefaction of the coal, a large portion of the effort here was directed to an examination of these conditions, seeking to maximize conversion to the gas oil boiling range and the extent of catalytic hydrodenitrogenation.

0-8412-0456-X/79/33-179-085$05.75/1
© 1979 American Chemical Society

With regard to basic nitrogen in the product of coal hydroliquefaction, an advantage of catalytic over thermal procedures for the denitrogenation process to be discussed is the ability of the hydrorefining catalysts, cobalt molybdate (Co–Mo) or nickel molybdate (Ni–Mo) on alumina, to partially hydrogenate neutral and acidic nitrogen compounds during the course of the hydrodenitrogenation, thereby converting them to basic nitrogen compounds precipitatable with dry HCl (3). In the case of carbazoles, e.g.:

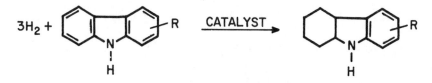

It was expected that the necessity of very deep removals of nitrogen during catalytic hydroliquefaction of coal would be obviated by the subsequent treating step to precipitate nitrogen compounds. In a sense it was proposed to use relatively mild hydroliquefaction conditions to achieve longer catalyst life by using the subsequent chemical separation of the nitrogen compounds. Similarly, the combination of mild hydroliquefaction conditions and HCl treatment as a trade-off for severe hydroliquefaction conditions to achieve the required degree of removal of nitrogen was expected to allow shorter contact times and lower temperatures to be used. This was to result in reduced C_1 to C_3 gas makes as well as higher yields of liquid products. In essence, the chemical extraction step should free the overall refining process from complete dependence on maintaining the very highest and constant activity in the hydrogenation catalyst. As will be shown, the combination of catalytic hydroliquefaction and HCl treatment has given a product in good selectivity containing less than 0.1% nitrogen, a level which has not been achieved in our work by catalytic hydroliquefaction alone.

A proposed processing scheme, many portions of which have been examined in the laboratory, is shown in Figure 1. Following filtration of ash from the product of catalytic hydroliquefaction, treatment of the coal liquids with dry HCl at ambient pressure gives an amine hydrochloride precipitate, an HCl extract, which separates as a mobile, dense liquid from the hydrocarbon phase. Thermal decomposition of the HCl extract returns HCl to the extraction step and leaves a concentrate of basic nitrogen compounds for conversion to hydrogen by a partial combustion process (4); the hydrogen is used in the hydroliquefaction and subsequent catalytic hydrorefining steps. Distillation of the neutralized hydrocarbon phase (the raffinate) gives gasoline, gas oil, and 1000°F+ residue. The reason for recycling a portion of the gasoline distillate to

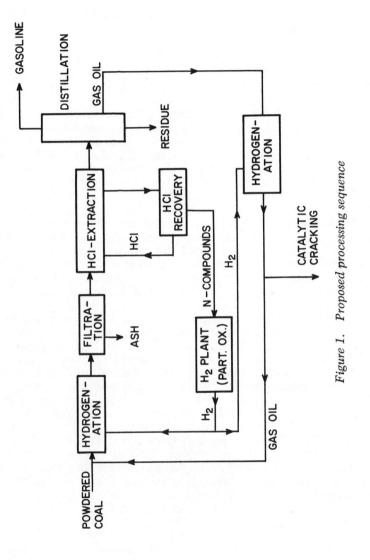

Figure 1. Proposed processing sequence

the HCl extraction step will become plain in the presentation of results below. Distillation bottoms may be recycled to the hydroliquefaction step for complete conversion to distillate product, while the gas oil requires hydrorefining prior to catalytic cracking to adjust the relative concentrations of saturates and aromatics as well as the distribution of polynuclear aromatic types in the latter. A portion of the gas-oil distillate is returned for use as solvent to the hydroliquefaction step.

Experimental

Apparatus. All hydroliquefactions were conducted in batch-stirred autoclaves, a 300-mL Hastelloy B reactor manufactured by Pressure Products Industries, Hatboro, Pa., and a 1-L, 316 stainless steel autoclave manufactured by Autoclave Engineers, Inc., Erie, Pa. The 300-mL reactor consisted of a fixed lower unit and a removable head equipped with a 1.5-in. diameter direct-drive turbine impellor, while the 1-L autoclave (Model AFP-1005) consisted of a fixed lower unit and a removable head equipped with a Magnedrive assembly to which a 1.25-in. turbine impellor was affixed. Both autoclaves were baffled and heated with appropriate electrical heating jackets. At the conclusion of a reaction the autoclave was cooled by means of air forced through an external cooling jacket. Each autoclave was provided with a thermowell, a gas inlet port connected by valving to hydrogen and nitrogen sources and to a pressure gauge, a port for connection to a rupture disk assembly, and a port for gas discharge.

Hydroliquefaction Procedure. In a standard hydroliquefaction experiment, the autoclave was charged with powdered coal (ca. 50 g in the 300-mL reactor and 167 g in the 1-L reactor), solvent (ca. 100 g in the 300-mL reactor and 333 g in the 1-L autoclave), and catalyst. After flushing with nitrogen, the reactor was charged with hydrogen, leak tested, and hydrogen sulfide was added to 200 psig at room temperature. Sulfiding of the catalyst was carried out by heating the stirred mixture to 200°C for 2 hr. Hydrogen sulfide was purged from the system by venting the cooled autoclave and alternately pressuring to 500 psig with hydrogen and venting to atmospheric pressure three times. Hydroliquefaction now was undertaken by adding hydrogen to 1000 psig to the reactor at room temperature. The temperature in the stirred autoclave then was raised to operating temperature as quickly as possible (ca. 1 hr at 425°C for the 300-mL reactor and 3 hr for the 1-L reactor), held at operating temperature for the desired length of time, and cooled quickly to room temperature with the aid of the external cooling jacket.

In the case of the 300-mL autoclave, hydrogen was added to bring the reactor pressure to 2600 psig when the desired reaction temperature was reached. Thereafter, the pressure in the stirred system was allowed to drop to 2400 psig at which time hydrogen was added again to raise the reactor pressure to 2600 psig. Hydrogen usage was calculated from the pressure drops, estimated free space, and application of the ideal gas laws. No provision was made for removal of gases from the 300-mL reactor during a reaction. To accomplish this the reactor was cooled after the desired reaction time and vented through a wet ice trap, a gas sampling bulb, and a wet test meter. If additional reaction time was desired, then fresh hydrogen was introduced, the stirred autoclave reheated, and the reaction continued.

The 1-L reactor was provided not only with a gas sampling valve which allowed for venting of the gas phase during a reaction, but also with a calibrated, 800-mL high-pressure hydrogen reservoir. During the course of a reaction all hydrogen was added to the 1-L reactor from the reservoir which was fitted with an accurate pressure gauge and thermocouple. The gauge was calibrated accurately so that the quantity of hydrogen added to the reactor from the reservoir was related precisely to the pressure drop and temperature of the reservoir. The amount of hydrogen added to the 1-L reactor was thus determined conveniently and precisely, regardless of the volume of reactants or reaction temperature. Furthermore, the gas sampling valve fitted to this autoclave permitted continual removal and analysis by mass spectrometry of gas from the reactor and replenishment with fresh hydrogen to maintain a high partial pressure of hydrogen during the reaction. Generally this was accomplished by allowing the reaction pressure to drop from 2600 to 2400 psig, venting briefly to 2300 psig, and then repressuring with hydrogen to 2600 psig. At the completion of a reaction, the gases were vented from the cooled reactor through a cold trap into a gas collection bag to permit thorough mixing. The gases then were expelled from the bag through a gas sampling bulb and a calibrated wet test meter and the gas sample was analyzed by mass spectrometry. This procedure allowed for the analysis of hydrogen, hydrocarbons, and hydrogen sulfide in the vent gases, but not for ammonia. Gas chromatography (GC) was used to analyze for nitrogen and carbon monoxide in the vent gases.

The slurry of liquid and solid reaction products was drained from the reactor and filtered under nitrogen pressure through Whatman No. 1 filter paper. All analyses of the filtered liquid product were performed on this filtrate which was not adulterated by washing solvents or changed in composition by heating to strip solvents. The solids were washed several times with a solvent (ca. 500 mL) composed of equal volumes of toluene and acetone. The liquid wash was filtered and stripped, and the liquid residue was weighed. The solids then were extracted in a Soxhlet thimble overnight with about a 10:1 weight ratio of toluene to wet filtered solids. The extract was stripped and weighed and the extracted solids were reported as "solids" after drying under vacuum. Overall material balances were within ±4%.

The filtered liquid products generally were analyzed for weight percent C, H, O, S, N_{Basic}, and N_{Total} (via Kjeldahl). Occasionally, molecular weight, $n - C_5$ insolubles, viscosity, and density determinations also were carried out.

Analytical Treatment of Hydrogenated Coal Liquids. SCHEME A. Filtered coal liquids were distilled in a standard vacuum distillation apparatus to recover the gasoline (IBP–400°F) and gas–oil (400°–1000°F) cuts (*see* Figure 2). The treatment with anhydrous HCl was applied to the composite of these cuts (IBP–1000°F) or to the gas–oil fraction, each diluted with an equal volume of *n*-hexane.

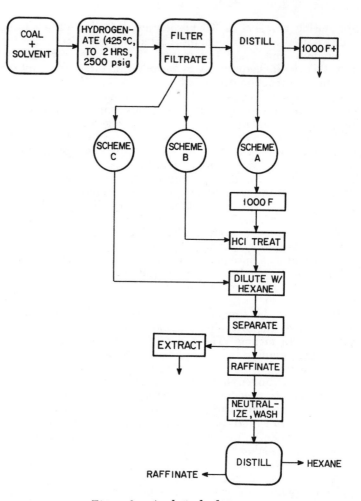

Figure 2. Analytical schemes

SCHEME B. Treatment with anhydrous HCl was applied to the entire filtered liquid product of coal hydroliquefaction diluted with an equal volume of *n*-hexane.

SCHEME C. The entire filtered liquid product of coal hydroliquefaction was mixed with an equal volume of *n*-hexane to obtain a precipitate of $n - C_6$ insolubles. It is worth noting that, at this point, further treatment of the raffinate with dry HCl to precipitate the remaining basic nitrogen compounds gave essentially the same overall weight of precipitate as Scheme B.

HCL TREATMENT OF COAL LIQUIDS. In a 100-mL centrifuge tube, 25.00 g of coal liquids are diluted to 50 mL with *n*-hexane, shaken thoroughly, and allowed to stand at room temperature for 16 hr. The precipitated $n - C_6$ insolubles are removed by centrifuging at 1500 rpm for 15 min and decanting and filtering the supernatant liquid through

No. 1 Whatman filter paper. Approximately 100 mL of $n - C_5$ are used to wash the precipitate. The latter is dried under nitrogen at 60°C, weighed, and reported as $n - C_6$ insolubles.

The supernatant liquid and $n - C_5$ washings are combined and saturated with anhydrous HCl at room temperature and atmospheric pressure for 5 minutes to precipitate a viscous, semisolid, nitrogenous dense phase. After standing for 16 hr at room temperature, the supernatant liquid is decanted and the precipitated material is washed with $n - C_5$, dried under nitrogen at 60°C, and weighed. The combined weights of C_6 insolubles and HCl extract are reported as HCl extract. The decanted liquids are concentrated to 75 mL over a water bath at 60°C and in a nitrogen atmosphere. After washing with 10% aqueous solutions of KCl, KOH, and KCl again, the liquid is topped by distillation to remove all C_6 diluent, weighed, and reported as percent raffinate. The raffinate is analyzed for C, H, O, N (by Kjeldahl), and S.

Coal. Illinois No. 6 coal from the Burning Star Mines was used in this study. It was obtained from Hydrocarbon Research, Inc., Trenton, New Jersey. It was crushed to ca. 70% through 100 mesh, dried to about 2% moisture, placed in plastic bags, and blanketed under nitrogen by HRI before shipment. Proximate and ultimate analyses of this coal appear in Table I.

Table I. Proximate and Ultimate Analyses of Illinois No. 6 Coal

| | Proximate Analysis | |
	As Received	Dry Basis
Moisture (wt%)	2.43	—
Ash (wt %)	10.44	10.70
Volatile (wt %)	36.38	37.29
Fixed Carbon (wt %)	50.75	52.01
Btu/#	12,326	12,633
Sulfur (wt %)	2.88	2.95

| | Ultimate Analysis | |
	Dry Basis (wt %)	MAF Basis (wt %)
Carbon	70.51	78.96
Hydrogen	4.96	5.55
Nitrogen	1.14	1.28
Chlorine	0.08	0.09
Sulfur	2.95	3.30
Ash	10.70	—
Oxygen (by difference)	9.66	10.82
H/C atomic Ratio	0.84	0.84

Solvent. The solvent used in most of this work was No. 24CA creosote oil from the Semet–Solvay Division of Allied Chemical Co., Morristown, New Jersey. The analysis of this material appears in Table II.

Table II.   Analysis of Creosote Oil

Carbon (wt %)	92.15
Hydrogen (wt %)	5.04
H/C atomic ratio (wt %)	0.66
Oxygen (wt %)	1.49
Sulfur (wt %)	0.71
Nitrogen (wt %)	0.88
Basic nitrogen (wt %)	0.49
Molecular weight	201

Distillation Data, VBR

Distilled (vol %)	(°F)
IBP	462
5	491
10	523
20	573
30	606
40	636
50	661
60	682
70	707
80	739
90	815
95	885
Final 97	897

Catalysts.   A commercially available catalyst, Filtrol HPC-5, comprised of cobalt molybdate on alumina in the form of the 1/16 in. extrudate was used as it was received. Two catalysts consisting of 1% CoO–2% MoO_3-on-bauxite differing only in the mesh size of the support, 8–20 and 100–270 mesh, were prepared in-house by dipping the granular and powdered bauxite supports in an aqueous solution of ammonium molybdate, cobaltous acetate, and 30% hydrogen peroxide. Prior to dipping, the supports were preignited at 500°–520°C in air for 16 hr and cooled to room temperature in a moisture-free atmosphere. After dipping, the solids were dried at 150°C in a circulating air oven for 16 hr and fired in a muffle furnace through which a slow stream of air was passed. The temperature during the firing was raised slowly from 200° to 500°C over a period of 5.5 hr and then maintained at 500°C for 16 hr.

Table III. Analyses

	Wt %	
	C	H
Raw Anthracene Oils	92.15	5.04
Hydrogenated Anthracene Oils	90.14	9.03

Results

Studies of Process Variables of Catalytic Hydroliquefaction. A factor of prime importance in the catalytic hydroliquefaction of coal is the H/C atomic ratio of the solvent. Table III lists the analyses of raw and hydrogenated anthracene oil solvents. By the time this degree of hydrogenation of the anthracene oil had been achieved, the rate of hydrogen uptake had slowed considerably. Table IV presents the experimental conditions and results of two similar hydroliquefaction experiments in the 300-mL stirred autoclave for prolonged reaction times at the relatively low temperature of 375°C in which raw and hydrogenated anthracene oil solvents were used. The effect of H/C atomic ratio of the solvent on the rate and extent of hydrogen uptake during catalytic coal hydroliquefaction is shown in Figure 3. Prehydrogenation of the solvent resulted in greatly enhanced removals of nitrogen and sulfur in the total liquid product and substantially reduced consumption of hydrogen during the hydroliquefaction step.

In a series of experiments at 425°C (Table V) in the 1-L stirred reactor at contact times greatly reduced from those used in the hydroliquefactions in the 300-mL reactor, the effects on total nitrogen in the filtered liquid product of reaction time, (*see* Experiments 3 and 4, Table V), the amount of catalyst, (*see* Experiments 4, 5, and 6, Table V), and the H/C atomic ratio of the solvent are brought out (*see* Experiments 4 and 7, Table V). Conversions of coal in all cases were close to 90%. These experiments were all carried out with the 8–20 mesh, granular, 1% CoO–2% MoO$_3$-on-bauxite in-house preparation of catalyst, whereas use of the 100–270 mesh powdered catalyst in an otherwise comparable experiment gave significant further reduction in total nitrogen (*see* Experiments 4 and 8, Table V). From Table V it is seen that the particle size of the catalyst exerts the largest effect on hydrodenitrogenation of the total coal liquids. These results are to be compared with those obtained using the commercially available HPC-5 catalysts (*see* Experiment 9, Table V). The bauxite-based catalysts are expected to be considerably less expensive than the alumina-based, commercially available catalyst.

of Anthracene Oils

	Wt %				
O	N$_T$	N$_B$	S	H/C	Mol Wt
1.49	0.88	0.49	0.71	0.66	201
0.54	0.34	0.05	0.30	1.20	208

Table IV. Effects of Solvent Hydrogen Content in

Experiment Number	Conditions (Time/min)
1	402/231
2	408/341

Analyses of Filtered Liquid Products

	Elemental Analyses (wt %)			
Experiment Number	C	H	O	N_T
1	90.59	7.61		0.64
2	89.68	9.25	0.30	0.07
Charge (calculated)				
1 (Anthracene Oil)	88.19	5.19	4.29	1.00
2 (Hydrogenated Anthracene Oil)	86.78	7.98	3.61	0.62

[a] Temperature 375°C; pressure 2500 ± 100 psig; 300-mL stirred reactor; fiiltrol HPC-5 catalyst.

Table V. Effects of Reaction Conditions and Catalysts on

	Experiment No.	
	3	4
Reaction conditions		
Time (min)	60	120
Catalyst (wt %)	10	10
H_2 absorbed MAF coal (wt %)	3.3	3.8
Reaction product data		
Filtered liquid product N (wt %)	0.62[b]	0.51[b]
1000°F− distillate (vol %)	93.0	95.0
1000°F− distillate N (wt %)	0.42	0.44
1000°F+ residue (vol %)	7.0	5.0

[a] 2500 psig; 425°C.
[b] 0.64 N (wt %) in charge (calculated).

Catalytic Hydrogenation of Coal–Anthracene Oil Slurries[a]

Recovery (wt %)	Conversion (wt %)	Hydrogen in Vented Gas (mol %)
96.3	87.2	75.6/75.8
98.7	90.7	57.8/90.2

Analyses of Filtered Liquid Products

Elemental Analyses (wt %)				Kin. Vis. (cSt)	
N_B	S	H/C	Mol wt	100°F	210°F
0.32	0.20	1.01	232	14.6	6.6
0.09	< 0.06	1.24	228	7.7	2.0
	1.49	0.71			
	1.20	1.10			

the Hydroliquefaction of Coal in Anthracene Oil Solvent[a]

		Experiment No.		
5	6	7	8	9
120	120	120	120	120
5	20	10	10	10
4.3	4.7	5.2	4.7	5.0
0.52[b]	0.43[b]	0.68[c]	0.44[b]	0.33[b]
93.0	94.0	91.0	92.6	94.0
0.45	0.35	0.55	0.35	0.27
7.0	6.0	9.0	7.4	6.0

[c] 0.98 N (wt %) in charge (calculated).

Table VI.   Elemental Analyses of Total and Distillate Products of

	Vol % of Total Liquid Product	Elemental Analyses (wt %)	
		C	H
Charge (calculated)		85.54	7.79
Experiment No. 3		89.13	8.25
Gasoline (Initial–400°F)	11		
Gas Oil (400°–1000°F)	82	89.78	8.17
Initial–1000°F	93	89.41	8.42
Bottoms	7	88.71	6.25
Experiment No. 4		89.96	8.10
Gasoline	22		
Gas oil	73	90.02	8.03
Initial–1000°F	95	90.45	8.39
Bottoms	5	86.79	5.85
Experiment No. 5		89.37	8.11
Gasoline	5		
Gas oil	88	89.67	10.00
Initial–1000°F	93.0	89.59	8.24
Bottoms	7.0		
Experiment No. 6		90.01	8.36
Gasoline	12		
Gas oil	82	90.10	8.36
Initial–1000°F	94	89.36	8.23
Bottoms	6	88.59	6.17
Experiment No. 9		90.38	8.76
Gasoline	11		
Gas oil	83		
Initial–1000°F	94	90.23	8.84
Bottoms	6	91.13	6.14

Catalytic Hydroliquefaction of Coal in a Solvent of Anthracene Oil

Elemental Analyses (wt %)				H/C Atomic Ratio	Specific Gravity 60/60°F
O	N_T	N_B	S		
3.93	0.64		1.29	1.08	
1.62	0.62			1.11	
	0.27				0.9016
1.31	0.46			1.09	1.0643
1.20	0.42	0.28	0.02	1.13	
1.40	1.53			0.85	
1.12	0.51			1.08	
	0.29				0.9203
1.22	0.46	0.28	0.036	1.07	1.0724
1.20	0.44	0.28	0.06	1.11	
1.03	1.49		0.58	0.81	
1.56	0.52	0.35		1.09	
				1.33??	
1.41	0.45	0.11		1.10	1.045
0.94	0.43	0.11		1.11	1.0545
	0.22				0.8826
0.87	0.36	0.17	< 0.04	1.11	1.0545
1.08	0.34	0.24	0.07	1.11	
0.78	1.34		0.06	0.84	
0.63	0.33	0.19	0.23	1.16	1.0329
	0.11				0.8392
	0.31				1.0455
0.86	0.30			1.18	
	1.10			0.80	

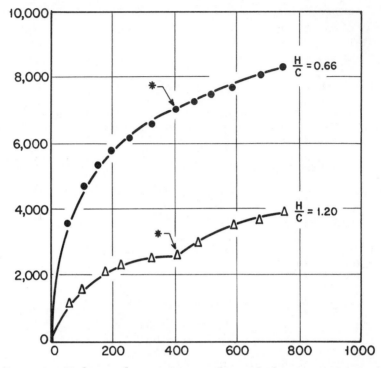

Figure 3. Hydrogen absorption rates: effect of hydrogen content of solvent. The asterisk indicates where the reaction was stopped and the reactor was vented and repressured with fresh hydrogen.

In all of these experiments conversions of coal to liquid products were 90 ± 3 wt %, recoveries of gases, liquids, and solids were greater than 93 wt % of the materials charged, and conversions of coal to 1016°F— distillable products were greater than 80 wt %. Distillation data and nitrogen content of distillate and residual fractions of the filtered liquid products from a number of experiments in Table V are presented in Table VI.

HCl Treating. The 1000°F— distillate fractions of the filtered liquid products from Experiments 3, 4, and 6 in Table V, upon saturation at ambient temperature and pressure with dry HCl gas, formed small amounts of dense, black liquid precipitates in each case (Table VII). A selective precipitation of nitrogen compounds took place as evidenced by the reduced nitrogen content in the raffinate phases and the increased nitrogen levels in the HCl extract phases. Of the nitrogen originally present in the distillate samples, the bulk remained in the raffinates, while minor proportions appeared in the extract phases and as water-soluble material. The possibility that solubility of the amine hydrochloride in the highly aromatic and oxygen-containing raffinate phases was causing the relatively constant levels of nitrogen in the raffinate phases prompted an examination of the effects of dilution with hexane on the yields of the HCl precipitates. The three schemes of analysis used are detailed in the Experimental Section. These analytical procedures were applied to the filtered liquid products of the experiments listed in Table V.

Table VIII lists the results of applying analytical Scheme A, B, and/ or C to Experiments 3, 4, 5, 6, 7, and 9 and to samples of raw and hydrogenated anthracene oil. In Scheme C the raw anthracene oil produced 15.4 wt % of precipitate, ostensibly asphaltenes, whereas the hydrogenated material gave no precipitate when diluted with hexane. However, hydrogenated anthracene oil in Scheme A formed 3.4 wt % of HCl extract. The effect of increasing severity of conditions during hydroliquefaction in proceeding from Experiment 3 to Experiment 6 in Table VIII is evident in the large decrease in $n - C_6$ insolubles in Scheme C. This is reflected in applying Scheme A to the 1000°F− distillates as well as Scheme B to the total filtered liquid products. Experiment 4, in which intermediate conditions of severity were used in hydroliquefaction, gave an intermediate yield of HCl extract when subjected to Scheme A. The poorest result in Scheme A of all the experiments was obtained in Experiment 7 in which partially hydrogenated anthracene oil with a H/C atomic ratio of 0.96 was used as the solvent. This corroborates the overall poor results obtained with this type of solvent in the experiments in the 300-mL stirred autoclave mentioned earlier.

In Table IX the effects of reducing catalyst particle size from 8–20 mesh to 100–270 mesh powder are evident. In Experiment 10, Scheme B produced significantly less HCl extract than was found using twice as much 8–20 mesh granular catalyst in Experiment 6 of Table VIII. Also

Table VII. Extraction of Nitrogen Compounds from 1000°F−
Distillate Fractions of Filtered Liquid Products in
the Absence of Hexane Diluent

	Exp. No. 3	*Exp. No. 4*	*Exp. No. 6*
Nitrogen analyses (wt %)			
1000°F− distillate	0.42	0.44	4.35
Raffinate	0.28 (52.8) [a]	0.25 (73.6)	0.25 (82.3)
Extract	2.59 (29.7)	2.20 (18.8)	2.47 (8.8)
Water solubles	0.0065 (18.5)	0.0042 (7.6)	0.0037 (8.7)
HCl precipitate (wt %)	6	3	1
N recovery (%)	122.7	98.9	86.5
Hydrocarbon recovery (%)	105.9	105.6	96.9

[a] Numbers in parentheses are normalized distributions of nitrogen.

Table VIII.   Precipitation of Nitrogen Compounds by Application tion of Coal in

Granular Catalysts Used

| | Exp. No. 3 | |
	Nitrogen Content (wt %)	Yield (wt %)
Scheme A products		
Raffinate	0.09	76.4
Extract	2.09	19.5
Loss		4.1
Coal as extract		65
Scheme B products		
Raffinate	0.08	64.8
Extract	1.50	32.2
Loss		3.0
Coal as extract		107
Scheme C products		
Raffinate	0.49	85.0
Extract	1.36	15.0
Loss		—
Coal as extract		52

| | Exp. No. 7 | |
	Nitrogen Content (wt %)	Yield (wt %)
Scheme A products		
Raffinate	0.10	75.6
Extract	2.80	23.7
Loss		0.7
Coal as extract		80
Scheme B products		
Raffinate		
Extract		
Loss		
Coal as extract		
Scheme C products		
Raffinate		
Extract		
Loss		
Coal as extract		

of Schemes A, B, and C to Products of Catalytic Hydroliquefac-Anthracene Oil

Granular Catalysts Used

Exp. No. 4		Exp. No. 5		Exp. No. 6	
Nitrogen Content (wt %)	Yield (wt %)	Nitrogen Content (wt %)	Yield (wt %)	Nitrogen Content (wt %)	Yield (wt %)
0.08	81.1	0.10	76.5	0.10	82.3
2.19	16.6	2.30	17.0		11.9
	2.3		6.5	2.53	5.7
	53		57		37
					73.3
					25.6
					1.1
					82.0
					96.6
					3.4
					—
					12

Exp. No. 9		Anthracene		Hydrogenated Anthracene Oil	
Nitrogen Content (wt %)	Yield (wt %)	Nitrogen Content (wt %)	Yield (wt %)	Nitrogen Content (wt %)	Yield (wt %)
0.10	83.3				91.9
2.31	8.4				3.4
	8.3				4.7
	24				n.a.
0.06	72.4				
2.21	15.9				
	11.7				
	54				
	92.2	0.61	84.6		
1.15	7.8	2.66	15.4		
	—		—		
	27		n.a.		

Table IX. Precipitation of Nitrogen Compounds by Application of
Schemes A, B, and C to Products of Catalytic Hydroliquefaction of Coal in Anthracene Oil

Powdered Catalyst Used

	Exp. No. 8		Exp. No. 10	
	Nitrogen Content (wt %)	Yield (wt %)	Nitrogen Content (wt %)	Yield (wt %)
Scheme A Products				
Raffinate	0.09	91.0		
Extract	3.65	9.0		
Loss		—		
Coal as extract		30		
Scheme B Products				
Raffinate			0.14	74.5
Extract			2.34	19.2
Loss				6.3
Coal as extract				63
Scheme C Products				
Raffinate				97.7
Extract			1.48	2.3
Loss				—
Coal as extract				

in Scheme A, use of 10% of the powdered catalyst in Experiment 8 of
Table IX produced a smaller yield of HCl extract from the 1000°F—
distillate product than was obtained with twice the amount of granular
catalyst in Experiment 6 of Table VIII.

The weight percents of coal converted to hexane insolubles in Scheme
C and to HCl precipitates in Schemes A and B are listed in Tables VIII
and IX. The extract values given for Schemes A and B have been corrected
for the HCl extract found in the liquefaction solvent. Elemental analyses
of various extract and raffinate phases produced by applying Schemes A,
B, and C in Table VIII are presented in Table X.

Discussion

That the combination of dilution with hexane and HCl treating is
more effective than dilution with hexane alone for precipitating nitrogen
compounds in the product of hydroliquefaction is seen in Experiment 3,
from the nitrogen analyses and yields of Schemes B and C extracts in
Table VIII. In this experiment under conditions of relatively short contact time, the yield of extract in Scheme B products is close to that of the
coal charged, indicating that essentially all of the liquefied coal is asphal-

tenic and/or HCl reactive, basic, and nonhydrocarbon. This situation is moderately improved in Experiment 6 in which both the reaction time and the amount of catalyst during the coal hydroliquefaction were twice the amount used in Experiment 3. Significant improvement regarding this was obtained in Experiments 8 and 10 in Table IX where, instead of 20 wt % of 8–20 mesh granular catalyst, 10 wt % of the identical catalyst powdered from −100 to +270 mesh size was used (all other conditions remaining essentially unchanged). In Experiment 10 the yield of HCl extract in Scheme B decreased to 19.2 wt %, indicating that approximately 37% of the coal had been converted to HCl unreactive nonbasic material —probably largely hydrocarbon.

Since, in most of these experiments, the raffinates from Schemes A and B contained 0.1 wt % or less of total nitrogen, the objective of converting coal to product containing 0.1% or less of nitrogen for possible application in catalytic cracking seems to have been attained.

For all the experiments, the amounts of coal precipitated as extracts in Schemes A, B, and C are related in Figure 4 with hydrogen consumptions during hydroliquefaction. At a consumption of approximately 5

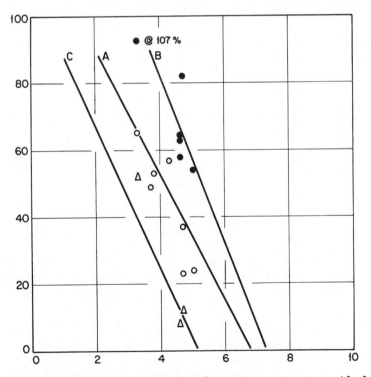

Figure 4. Amounts of coal precipitated as extracts as functions of hydrogen absorbed: (○), Scheme A; (●), Scheme B; and (△), Scheme C.

Table X. Elemental Analyses of Extract and Raffinate Phases
Hydroliquefaction of Coal in

	Elemental Analyses (wt %)		
	C	H	O
Anthracene Oil			
Raffinate (Scheme C)	92.01	5.82	1.18
Extract (Scheme C)			
Experiment No. 3			
Raffinate (Scheme A)	90.60	8.32	0.94
Raffinate (Scheme B)	90.46	8.36	0.78
Raffinate (Scheme C)			
Extract (Scheme A)			
Extract (Scheme B)			
Extract (Scheme C)			
Experiment No. 4			
Raffinate (Scheme A)	90.65	8.24	0.94
Extract (Scheme A)			
Experiment No. 5			
Raffinate (Scheme A)	89.37	8.14	0.84
Extract (Scheme A)			
Experiment No. 6			
Raffinate (Scheme A)	90.02	8.20	0.70
Raffinate (Scheme B)	90.60	8.33	
Extract (Scheme A)			
Extract (Scheme B)	84.58	7.00	1.75
Experiment No. 7			
Raffinate (Scheme A)	91.46	7.53	
Extract (Scheme A)			1.54
Experiment No. 9			
Raffinate (Scheme A)	90.27	8.91	0.59
Raffinate (Scheme B)	90.78	8.63	0.59
Extract (Scheme A)	86.31	6.91	2.14
Extract (Scheme B)			

wt % of hydrogen based on maf coal, precipitate is no longer formed
according to Scheme C in diluting the total filtered liquid product with
hexane. However, at 5 wt % consumption of hydrogen about 60 wt %
of the dissolved coal is precipitated as HCl extract in Scheme B and 35
wt % in Scheme A. Furthermore, Figure 3 shows that the projected
requirement of hydrogen under the conditions of these experiments, which
is to completely suppress precipitation of HCl extract by Scheme B, is
ca. 7.3 wt % based on maf coal and about 6.8 wt % from Scheme A.

Obtained by Applying Schemes A, B, and C to Products of Catalytic a Solvent of Anthracene Oil

Elemental Analyses (wt %)				H/C Atomic Ratio
N_T	N_B	S	Cl	
0.61				0.75
2.66				
0.09	0.00	0.08		1.10
0.08		0.09		1.11
0.49				
2.09				
1.50				
1.36				
0.08	188 ppm	< 0.06		1.09
2.19				
0.10				1.09
2.30				
0.10				1.09
		0.05	137 ppm	1.10
2.53				
1.80		< 0.06		0.99
0.10		0.10		1.00
2.80				
0.10				1.18
0.06				1.13
2.31				0.95
2.21				

Thus, in the simultaneous liquefaction and primary catalytic hydro-refining of coal as practiced here, at the point of complete disappearance of hexane-insoluble asphaltenes very large proportions of the original coal remain as nonhydrocarbons in both the total filtered liquid product and in the distillate boiling below 1000°F. At this stage of hydrorefining, an additional input of ca. 2.2 wt % of hydrogen is required to convert all HCl-reactive, basic nonhydrocarbons to hydrocarbons and neutral nonhydrocarbons in the total filtered liquid product and about 1.7% in the 1000°F— distillate. It is interesting that of the hydrogen required to liquefy coal and to convert the liquefied product completely to hydrocarbons and neutral nonhydrocarbons, approximately 70% is needed for liquefaction and conversion of the hexane-insoluble asphaltenes to hexane-soluble, basic nonhydrocarbons, whereas 30% is required to convert the latter to hydrocarbons and neutral nonhydrocarbons.

Consumptions of about 7 wt % of hydrogen during the simultaneous liquefaction and primary catalytic hydrogenation of coal under the conditions and catalysts specified will yield coal liquid highly upgraded with regard to nitrogen and asphaltene contents. However, a parallel investigation of concentration of aromatics and distribution of aromatic types in products of this sort is required, since these factors also critically influence the quality of catalytic cracking charge stock.

Conclusions

The experimental results in this chapter show that potential charge stock for catalytic cracking in the gas oil boiling range containing less than 0.1% nitrogen can be obtained from the product of catalytic hydroliquefaction of Illinois No. 6 coal by HCl treatment. However, it is also clear from these results that the yields of HCl-precipitated nitrogen compounds are excessive and represent a serious loss in yield of the desired hydrocarbon product. More severe conditions during the hydroliquefaction step, such as higher pressures of hydrogen and/or higher temperatures to increase the extent of hydrodenitrogenation, are required to correct this deficiency.

In order to achieve a balance between the depth of hydrodenitrogenation needed during hydroliquefaction and the yield of amines recovered for generating hydrogen, a determination of the hydrogen-producing potential of the amine extract must be made. From a practical point of view this requires the prior development of a process for the thermal decomposition of the amine hydrochloride precipitate quantitatively to separate and recover both hydrogen chloride and the amine concentrate. From a complete set of inspection data and elemental analyses on the amine concentrate a calculation can be made of its hydrogen-producing potential in a partial combustion process such as the Texaco partial oxidation process (4).

Acknowledgment

The research reported here was supported by the U.S. Department of Energy under contract EF-76-C-01-2306.

Literature Cited

1. Johnson, C. D.; Chervenak, M. C.; Johanson, E. S.; Stotler, H. H.; Winter, O.; Wolk, R. H. "Present Status of the H–Coal Process," Clean Fuel from Coal Symposium, Institute of Gas Technology, Chicago, Illinois, September 10, 1973.
2. Sternberg, H. W.; Raymond, R.; Schweighardt, F. K. *Science* **1975,** *188,* 49.
3. Ahmed, M. M.; Crynes, B. L. *Prepr. Div. Pet. Chem., Am. Chem. Soc. 22,* No. 3, 971 (1977).
4. Shimizu, Y. *Chem. Econ. Eng. Rev.* **1978,** 7, 9.

RECEIVED June 28, 1978.

Upgrading Primary Coal Liquids by Hydrotreatment

A. J. deROSSET, GIM TAN, and J. G. GATSIS

Corporate Research Center, UOP Inc., Des Plaines, IL 60016

Primary coal liquefaction products from three processes—solvent-refined coal, Synthoil, and H–Coal—were hydrotreated. Upgrading was measured in terms of the decrease in heptane and benzene insolubles, the decrease in sulfur, nitrogen, and oxygen, and the increase in hydrogen content. Hydrotreating substantially eliminated benzene insolubles and sulfur. An 85% conversion of heptane insolubles and an 80% conversion of nitrogen was obtained. Catalyst stability was affected by metals and particulates in the feedstocks.

Primary coal liquids are high-viscosity black oils or pitches. Compared with petroleum crudes, they are deficient in hydrogen, only partially soluble in benzene, and contain relatively high levels of oxygen and nitrogen. They contain a variable amount of ash and unconverted coal, depending on the efficiency of the mechanical removal of solids achieved in each process.

These coal liquids originally were intended to serve as boiler fuels. The sulfur content may be sufficiently low to qaulify them as replacement for high-sulfur coal, and bring sulfur dioxide emissions within acceptable limits. However, for large boilers, such as major base load power plants, pollution control via coal hydroliquefaction probably is not competitive with stack gas scrubbing (1).

In other fuel markets, coal liquids can be more competitive. Industrial boilers presently are not amenable to stack gas scrubbing. The same is true of smaller utility plants. In particular, peak load units require a clean, storable liquid fuel as an alternative to natural gas. However, the high viscosity of primary coal liquefaction products is undesirable for many of these applications. Also, their residual sulfur and nitrogen contents may be excessive as emission standards become more stringent.

0-8412-0456-X/79/33-179-109$05.00/1

If there is a petroleum shortage in the future the problem will be to provide a portion of the demand for transportation fuels from coal. Transportation fuels—jet, diesel, and gasoline—are distillates boiling below 650°F. However, the quantity of such distillates in primary coal liquids is small.

Primary coal liquids must be upgraded in order to serve these markets. A logical route is to use current black oil conversion technology as practiced in the petroleum industry (2). An applicable UOP process is RCD Unibon (3). This comprises the direct processing of petroleum residues to reduce the sulfur and nitrogen content of heavy fuel oil or to combine desulfurization with conversion of residue to lighter, more valuable products.

This chapter reports results of applying a catalytic hydrorefining process to four coal liquids: solvent-refined coal (SRC) process filter feed, SRC extract product, Synthoil, and H–Coal process hydroclone underflow. The achieved upgrading is evaluated in terms of reduction in benzene and heptane insolubles, reduction in sulfur, nitrogen, and oxygen, an increase in hydrogen content, and a yield of lower boiling products.

Hydrotreatment was carried out over a commercial UOP black oil conversion catalyst in bench-scale units of 200–800 mL catalyst capacity. Temperature range was 375°–450°C and the pressure range was 2000–3000 psig. Weight hourly space velocity (WHSV) varied from 0.1 to 1.0 depending on the heptane-insoluble content of the feed. A flow diagram of a typical plant is shown in Figure 1. The stripper bottoms usually

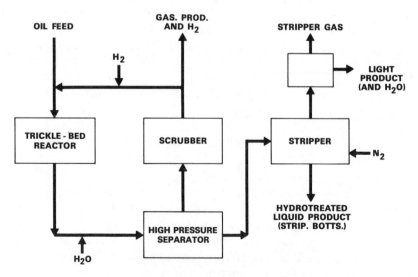

Figure 1. Hydrotreatment plant

comprised over 90% of the exit streams. It was the only portion analyzed in detail and is referred to as product.

The first stock run was SRC filter feed obtained from the Pittsburg and Midway Coal Company SRC pilot plant at Ft. Lewis, Washington. This contains all of the recycle process solvent, ash, and unconverted coal. The stock was filtered prior to hydrotreating. (Table I compares inspections of the filter feed and filtrate.) Filtration upgraded the oil portion of the stock.

Table I. Inspection of SRC Stocks

	SRC Filter Feed	UOP Filtrate	SRC
°API @ 60°F	−11.8	−5.8	−13.7
Distillation (ASTM D-1160)			
IBP (°F)	400	400	648
10%	579	550	905
30%	654	620	—
50%	729	685	—
70%	925	825	—
SRC (850°F+) (Vol %)	30.3[a]	29.0	93.0
Heptane insoluble (Wt %)	29.7[a]	26.9	89.4
Benzene insoluble (Wt %)	16.7[a]	9.4	34.7
DMF/xylene insoluble (Wt %)[b]	6.7	—	0.15
Ash (ASTM) (Wt %)	4.0	0.01	0.11

[a] Corrected to "ash plus unconverted coal" free basis.
[b] Assumed to be "ash plus unconverted coal."

The dimethylformamide (DMF)/xylene insolubles, assumed to represent ash and unconverted coal, were eliminated substantially. The benzene and heptane insolubles in the feed, adjusted to an "ash plus unconverted coal" free basis, were reduced substantially and the ASTM ash dropped from 4% to 0.1%. The SRC content of the filtrate, defined as percentage 850°F+ bottoms was 29%, as compared with 30.3% in the feed after correction to an "ash plus unconverted coal" basis.

Hydrotreating the filtrate resulted in the upgrading shown in Table II. Generally the degree of upgrading corresponded to the severity of the operation. For example, Product B was obtained at a space velocity twice that used to make Product A. Similarly, within the limits of the range of experimental conditions used, upgrading improved with increasing temperature and pressure. No substantial decline in performance was observed over 144 hr of operation.

Table II. Upgrading Primary

	SRC Filter Feed[a]		
	Feed	Product A	Product B[b]
Heptane insoluble (Wt %)	26.9	0.5	3.1
Benzene insoluble (Wt %)	9.4	0.03	0.08
Sulfur (Wt %)	0.72	< 0.02	0.06
Nitrogen (Wt %)	1.28	0.14	0.42
Oxygen (Wt %)	3.81	0.25	1.88
Hydrogen (Wt %)	6.90	10.33	9.46

[a] Filtered by UOP.
[b] Obtained at twice the space velocity used to make Product A.

Of the 29% SRC contained in the feed, a considerable amount was converted to lower boiling materials. However, an unambiguous product distribution is difficult to establish because of the preponderance of process solvent in the feed. Therefore, SRC was hydrotreated without addition of extraneous solvent.

Table I shows the properties of the SRC used. This material is produced in the SRC pilot plant by filtration of the solvation reactor effluent and vacuum stripping of process solvent. It contained 93% SRC and 0.11% ASTM ash. For comparison, the ash content of the UOP-filtered SRC filter feed was 0.01%, or 0.03% based on SRC.

In order to process SRC in the pilot plant equipment without using an extraneous solvent, a portion of the hydrotreated product was recycled to the feed reservoir as diluent. Recycle was essentially internal, and SRC and hydrogen were the only net feeds to the system. Pulverized SRC was added to the hot reservoir at a ratio of two parts SRC to three parts recycled product (combined feed ratio = 2.5). Vigorous stirring ensured that no undissolved solids were taken into the feed pump system.

A small amount of start-up solvent was required. This was generated by batch autoclave runs, using conditions and a catalyst comparable with those used in the subsequent continuous run. Catalyst was separated by filtration, and benzene extracted to yield additional oil for start-up.

The continuous run was on stream for 456 hr when a 300-psig pressure drop developed across the reactor, necessitating shutdown. Analyses of the composite product are compared with those of the feed in Table II.

The heptane-insoluble content of the composite product was 9.76%, representing an 89% conversion. The variation of heptane insolubles with time is shown in Figure 2. Initial conversion of heptane insolubles was 94% while final conversion was 86%.

Coal Liquids by Hydrotreatment

SRC		Synthoil Centrifugate[e]		H–Coal Hydroclone Underflow[e]	
Feed	Product	Feed	Product	Feed	Product
89.4	9.76	18.4	3.9	49.0	3.45
34.7	1.68	4.4	0.4	29.0	0.06
0.78	0.02	0.56	0.02	0.66	< 0.02
2.12	0.55	1.31	0.38	1.30	0.17
2.71	0.20	2.27	0.35	4.38	0.39
5.63	8.76	7.42	9.47	6.96	9.56

[e] Filtered and water washed by UOP.

Over the 120–144-hr period, mass balance, chemical analysis, and fractionation gave the fuel product and hydrogen distributions shown in Table III. Of the hydrogen consumed, 79% was retained in the fuel products. The remainder was used to remove heteroatoms.

Table III. Distributions in Hydrotreatment of Pretreated Primary Coal Liquids

	Charge Stock		
	SRC	Synthoil Centrifu-gate	H–Coal Hydroclone Underflow
Pretreatment	None	Filtration water washing	Filtration
Fuel product (Wt % of feed)	98.7	99.0	96.5
Fuel product distribution (Wt %)			
gas (C_1–C_4)	1.5	1.7	2.5
gasoline (C_5—400°F)	5.2	9.0	1.9
kerosine (400°–527°F)	5.6	11.3	14.6
light gas oil (527°–752°F)	32.7	37.5	47.2
heavy gas oil (752°–850°F)	14.2	10.6	9.6
bottoms	40.8	29.9	24.2
Total	100.0	100.0	100.0
Hydrogen consumption SCF/B	3,790	2,000	2,760
Distribution of hydrogen consumption (Wt %)			
C_5+ liquid product	73.6	75.7	69.2
gas (C_1–C_4)	5.4	7.6	7.2
H_2O	7.4	8.5	15.0
H_2S	0.9	1.2	1.2
NH_3	7.6	7.0	7.4
Total	100.0	100.0	100.0

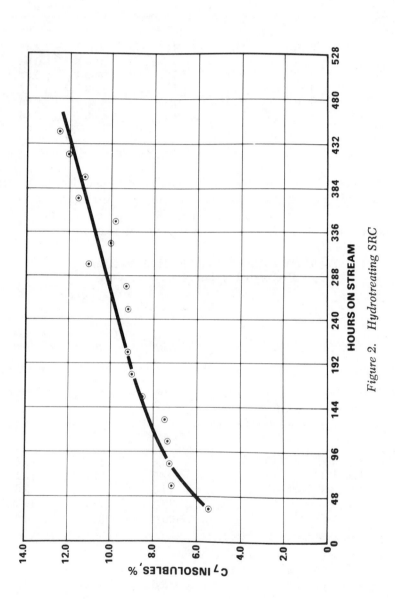

Figure 2. Hydrotreating SRC

Synthoil, produced at the Pittsburgh Energy Research Center, was a centrifuged product. It contained 1.4% ash and over 0.1% ammonium chloride, which necessitated both filtration and hot-water washing as pretreatments. Pertinent inspections are given in Table IV. The heptane and benzene insolubles were relatively low, less than those of SRC filter feed. The stock could be processed without recourse to product recycle.

Table IV. Inspection of Synthoil Stocks

	Synthoil Centrifugate	Filtrate	Washed Filtrate
°API @ 60°F	−5.7	−3.5	−4.3
Distillation (ASTM D-1160)			
IBP (°F)	395	407	455
10%	510	515	552
30%	642	635	665
50%	762	760	795
70%	945	951	970
Heptane insoluble (Wt %)	19.0	15.2	18.4
Benzene insoluble (Wt %)	6.3	5.0	4.4
DMF/xylene insoluble (Wt %)	3.0	—	—
Ash (ASTM) (Wt %)	1.40	0.02	0.015

A hydrotreating run of 436 hr was completed on pretreated Synthoil. Analyses of the composite product are compared with those of the feed in Table II.

The heptane-insoluble content of the composite product was 3.9%, representing 79% conversion. The variation of heptane insolubles with time is shown in Figure 3. After 80 hr the operation was reasonably stable.

Over a 102–126-hr period, mass balance, chemical analysis, and fractionation gave the fuel product and hydrogen distributions shown in Table III. Of the hydrogen consumed, 83% was retained in the fuel products.

H–Coal hydroclone underflow was supplied by Hydrocarbon Research, Inc. It contained 9.09% ash and required filtration. Because of the relatively high viscosity of the stock, filtration was difficult, and only a limited amount of hydrotreater feed was prepared. Inspections are given in Table V. Thermal and atmospheric exposure during filtration downgraded the stock. The benzene and heptane insolubles in the product were substantially higher than those in the feed, when corrected to an "ash plus unconverted coal" free basis. The filtrate still contained 0.12% ash.

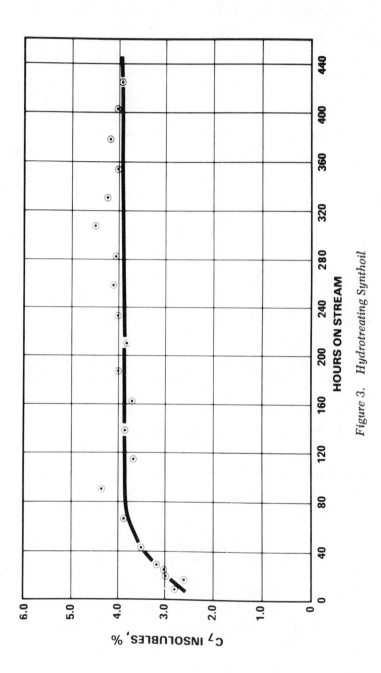

Figure 3.   Hydrotreating Synthoil

Table V. Inspection of H–Coal Stocks

	Hydroclone Underflow	Hydroclone Underflow Filtrate
°API @ 60°F	−16.5	−17.7
Distillation (ASTM D-1160)		
IBP (°F)	466	493
10%	560	567
30%	690	680
50%	876	822
Heptane insoluble (Wt %)	41.0[a]	49.0
Benzene insoluble (Wt %)	23.4[a]	29.0
DMF/xylene insoluble (Wt %)[b]	15.1	0.03
Ash (ASTM) (Wt %)	9.09	0.12

[a] Corrected to "ash plus unconverted coal" free basis.
[b] Assumed to be "ash plus converted coal."

A 178-hr hydrotreating run was made, during which the heptane-insoluble conversion dropped 10%. Results obtained at 54–78 hr on stream are given in Table II. Mass balance, chemical analysis, and fractionation gave the fuel product and hydrogen distributions shown in Table III.

The preceding data lead to a comparison of Synthoil and SRC stocks in two respects—their relative susceptibility to upgrading by hydro-refining and their relative effect on catalyst stability.

Table VI recapitulates the relative response of Synthoil and SRC to hydrotreating in terms of percentage conversion of heptane insolubles, as reported in Table II. Conversion of heptane insolubles was substantially greater in the case of SRC. While the Synthoil was processed at a higher space velocity, the throughput rate of Synthoil feed heptane insolubles was only half that of the SRC feed heptane insolubles. Temperature in the Synthoil run was 5°–10°C lower than in the SRC run.

Table VI. Relative Response of Synthoil and SRC to Upgrading by Hydrotreatment

	Heptane Insolubles in Feed (Wt %)	Relative Hourly Space Velocity	Relative Heptane-Insoluble Feed Rate	Heptane-Insoluble Conversion (Wt %)
SRC	89.4	1	1	86–96
Synthoil	18.4	2.5	0.5	79

This comparison is reinforced by a comparison of Synthoil with UOP-filtered SRC filter feed. Figure 4 illustrates the response of the two stocks to hydrotreating as a function of pressure and space velocity. The temperature range of the experiments was 9°C. Conversion of SRC heptane insolubles was first order up to 98%, and the effect of pressure was relatively small. Conversion of Synthoil heptane insolubles never exceeded 80% under comparable conditions, and it was strongly dependent on pressure.

The asphaltenes in SRC stocks are more responsive to hydrotreating than those in Synthoil. Whether this is because of differences in the two processes, or because of differences in the coals used to produce the particular materials studied, is a matter of speculation.

A recapitulation of the catalyst stability data reported indicates that within the time scale of the hydrotreating runs, the UOP-filtered SRC filter feed gave relatively stable performance. SRC itself caused substantial catalyst deactivation; Synthoil gave stable performance after an initial deactivation. Since dissolved metals and particulate matter are known to have an adverse effect on catalysts, a correlation was sought based on an analysis for these components.

Briefly, the analysis comprises dissolution of the coal liquid in DMF/xylene (2:1) and filtration through a 0.2-μm Fluoropore filter. The filtrate is analyzed for metals (Fe, Ti, Al, Ca, Mg, Na, and Si) by atomic absorption spectroscopy (AAS) to give soluble metals. The filter cake is

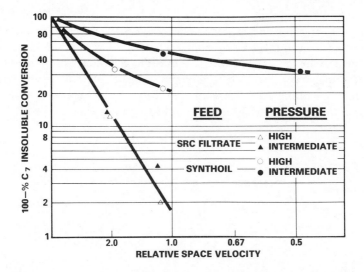

Figure 4.   Comparison of conversion of C_7 insolubles in SRC and Synthoil-derived stocks

Table VII.   Particulates and Soluble Metals in Hydrotreater Feedstocks

	Soluble Metals (ppm)		Particulates (ppm)	
	Total	Ti	Total	Metals (as oxides)
SRC/UOP filtrate	51	34	100	88
SRC	281	118	1,500	591
Synthoil	23	16	1,600	227

weighed to give particulates and is analyzed further by wet digestion and AAS to give insoluble metals. Results of analyses of the two SRC stocks and Synthoil are given in Table VII.

In all cases the predominant soluble metal was Ti. Some of the other metals, such as Si and Al, may actually be fine particulates passing through the filter.

The particulates in the SRC filter feed (after UOP filtration) appear to be predominately inorganic. The metals containing the particulates (calculated as oxides) account for almost all. In SRC, metal oxides could account for almost half of the particulates measured, especially if some fine particulates are reporting analytically to the filtrate instead of to the filter cake.

Particulates were mostly organic in Synthoil. Soluble metals are relatively low.

If deactivation of black-oil conversion catalysts is a result of pore plugging, two mechanisms are plausible. Pore plugging by dissolved metals proceeds by successive decomposition of molecules of a metal organic complex until a plug develops by accretion. Pore plugging by particulates is accomplished by the trapping of a single particle or a small number of particles in a pore of a size comparable with the particle diameter.

Both particulates and dissolved metals are probably involved in pore plugging when processing primary coal liquids. Analyses suggest that the former are more significant in the case of the Synthoil sample, and both are significant in the case of the SRC sample.

Literature Cited

1. Weiss, L. H., presented at the Eleventh Intersociety Energy Conversion Engineering Conference, State Line, NV, Sept. 12–17, 1976.
2. Ranney, M. W. "Desulfurization of Petroleum"; Noyes Data Corp.; Park Ridge, NJ, 1975.
3. Turnock, P. H., presented at the Meeting of the Japan Petroleum Institute, Tokyo, Japan, May 8–9, 1975.

RECEIVED June 28, 1978. The work was supported by the U.S. Energy Research and Development Administration under Contract E(49–18)2010.

8

Catalytic Hydroprocessing of Solvent-Refined Coal

E. N. GIVENS, M. A. COLLURA, R. W. SKINNER,
and E. J. GRESKOVICH

Corporate Research and Development Department, Air Products and
Chemicals, Inc., Allentown, PA 18105

*Authentic and synthetic solvent-refined coal filtrates were
processed upflow in hydrogen over three different commercially available catalysts. Residual ($>850°F$ bp) solvent-
refined coal versions up to 46 wt % were observed under
typical hydrotreating conditions on authentic filtrate over a
cobalt–molybdenum (Co–Mo) catalyst. A synthetic filtrate
comprised of creosote oil containing 52 wt % Tacoma
solvent-refined coals was used for evaluating nickel–molybdenum and nickel–tungsten catalysts. Nickel–molybdenum
on alumina catalyst converted more $850°F+$ solvent-refined
coals, consumed less hydrogen, and produced a better product distribution than nickel–tungsten on silica alumina. Net
solvent make was observed from both catalysts on synthetic
filtrate whereas a solvent loss was observed when authentic
filtrate was hydroprocessed. Products were characterized by
a number of analytical methods.*

D uring the period June, 1975 to October, 1977 Air Products and
Chemicals, Inc. studied the catalytic hydroprocessing of solvent-
refined coal (SRC) products. This work was sponsored by the United
States Energy Research and Development Administration under a contract entitled "The Chemical Characterization, Handling, and Refining
of Solvent-Refined Coal to Liquid Fuels." Results from that study are
the subject of this presentation. The final report recently was issued as
FE2003-27, September 1977, available through NTIS.

The program objectives were: (a) to characterize the physical and
chemical properties of SRC and its derivatives; (b) to develop tech-

0-8412-0456-X/79/33-179-121$06.00/1
© 1979 American Chemical Society

nology for the desulfurization, denitrogenation, and hydrocracking of SRC using commercially available catalysts; and (c) to determine the bulk handling characteristics of SRC.

The program was broken down into several distinct tasks in order to accomplish these objectives. First, a literature survey was performed to gather articles pertinent to the hydrotreating and hydrocracking of SRC as well as information on analytical methods for the characterization of SRC and its derivatives.

Task 2 involved the development of analytical techniques and capabilities, first to characterize the products of the SRC process and secondly to provide support for the hydroprocessing studies.

The third task, bulk-handling characteristics of SRC, was conducted by Lehigh University under subcontract to Air Products and Chemicals, Inc. This work is not discussed in this chapter.

Task 4 was the design and construction of the test unit which subsequently was used to carry out Tasks 5 and 6. These tasks involved the experimental work for the desulfurization, denitrogenation, and hydrocracking of SRC.

Experimental

The equipment, operating procedures, and analytical methods used in the program are mentioned briefly.

Equipment. In designing the apparatus to be used for the hydrotreating and hydrocracking of SRC, heavy reliance was placed on the petroleum-processing experience in the Houdry Division of Air Products. A continuous process development unit was designed and constructed to catalytically hydroprocess nominally 1 L/hr of SRC liquid. A simplified flowsheet of the unit appears in Figure 1.

Not shown in the flowsheet is a 60-gal, heated, stirred tank used to prepare and store feedstock. The feed flows from this mixing tank to the feed tank, which is an 8-L vessel. The liquid inventory in the feed tank was monitored with a differential pressure transmitter from which the liquid flow rate was determined. The feed tank holds enough liquid for one balance period up to 8 hr.

The feed pump was a positive-displacement, piston-type metering pump. All liquid transfer lines in the system were traced with electric resistance heaters which allowed us to process heavy SRC solutions.

Hydrogen gas was preheated and mixed with feed liquid before entering the reactor. The reactor effluent was cooled to transfer-line temperature in a heat exchanger and passed to a separator. Both liquid and gas streams were depressurized through separate control valves to 30 psig. The liquid stream is degassed at the lower pressure.

The two gas streams passed through a water-cooled condenser to remove water and light organic liquids. The remaining gas stream rate is measured by being passed through a flowmeter and analyzed in a process gas chromatograph (GC).

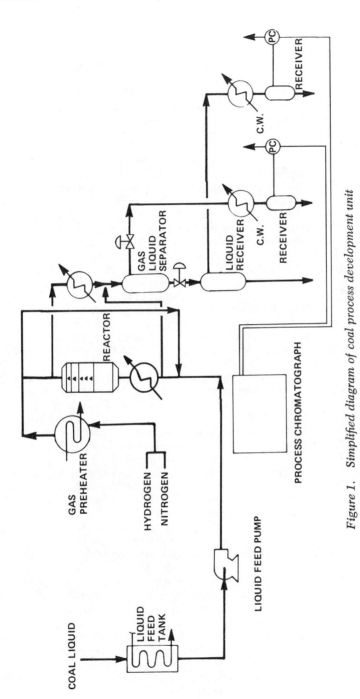

Figure 1.   Simplified diagram of coal process development unit

The liquid product in the main liquid receiver was combined with the organic phases in the condenser receivers. The two water layers were combined also.

The fixed-bed reactor has a 2-in. inside diameter and a 60-in. inside length for a volume of 3 L (Figure 2). It is constructed of A-286 high-temperature alloy which can be operated at 5000 psig at 1000°F. It can withstand temperature overruns up to 1500°F for short periods. A thermo-well runs through the center of the catalyst bed the entire length of the reactor. Heat is supplied to the reactor in five separately controlled heating zones.

Typically, the reactor was loaded with a preheat section in the bottom for an upflow configuration. The preheat section, which contained 1.5 L of Harshaw Tab Alumina rested on stainless steel wool. The 1-L catalyst bed was placed on top of the alumina preheat section. The small volume at the top of the reactor was filled with additional alumina topped with stainless steel wool to retain the catalyst in the reactor. The thermocouples were inserted at the designated locations in an axial thermowell which runs through the center of the catalyst bed.

The range of process variables which can be studied on this unit are: pressures to 3000 psig; reaction temperatures to 1000°F; transfer-line temperatures to 600°F; liquid feed rates from 0.3–3.8 L/hr; and gas feed rates from 50–1000 std L/hr.

Operation. Catalysts were pretreated by first reducing at 500°F under hydrogen for several hours. Presulfiding was accomplished with either an SRC filtrate or Allied Chemical 24 CB creosote oil over a 24-hr period. Allied 24 CB creosote oil contains 0.7 wt % sulfur. At a 10% sulfur removal level approximately 6 wt % sulfur on catalyst was generated. Each set of process conditions was maintained for several hours during which time several balances were taken. Runs lasted up to 180 hr. All runs discussed in this chapter were voluntarily shut down.

An exotherm of up to 30°F was observed across the reactor. Run temperatures reported in our discussion refer to the temperatures at the middle of the bed. The reaction was monitored by following the disappearance of heptane and benzene insolubles, removal of sulfur and nitrogen from the liquid phase, and hydrogen consumption across the reactor. As the reaction progressed the C_7 insolubles would appear to line out at about 30 hr on-stream.

Analytical. Gas streams were analyzed by an on-line process GC for the following: hydrocarbons; sulfur compounds—H_2S, COS; carbon oxides—CO, CO_2; ammonia; hydrogen; oxygen; and nitrogen. Water products were analyzed for nitrogen and sulfur.

Main receiver product was subjected to a thorough product work-up. Elemental analyses were performed first on total liquid product after which it was fractionated to give the following fractions: light distillate—Initial Boiling Point (IBP) to 420°F; solvent range distillate—420°F to final cut point; and residual fraction. Each fraction was subjected to elemental analysis. Various samples also were analyzed by the following methods: heptane and benzene solubility; gas–liquid chromatography (GLC); high-pressure liquid chromatography (HPLC); IR spectroscopy; and thermal gravimetric analysis.

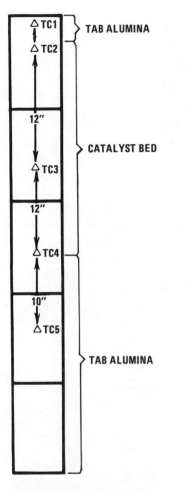

Figure 2. Reactor diagram

The solubility measurements were made using the UOP Method 614–68 entitled "Heptane-Insoluble Matter in Petroleum Oils Using a Membrane Filter: Heptane-Insoluble Method with Ultrasonic Agitation." The method description was kindly supplied by the Universal Oil Products Company.

Results and Discussion

Hydrotreating of Tacoma filtrate was investigated over both an inert, low-surface-area alumina and a silica-promoted cobalt–molybdena (Co–Mo) on alumina catalyst. The filtrate feed used in this study was from a run made at Tacoma on June 21, 1976 operating with Kentucky Colonial coal as feed at 843°F, 1500 psig at a solvent to coal ratio of 1.6:1. This information was kindly supplied by G. R. Pastor, Pittsburg and Midway Coal Company, Dupont, Washington. Recycle solvent upper cut point was apparently at 850°F as determined by our distillation data on the samples. Analytical data on the filtrate is presented in Table I.

Thermal Stability of Filtrate. The thermal stability of filtrate from the SRC process to typical process conditions for hydrorefining was tested in a 40-hr, noncatalytic hydrotreating run. Filtrate was passed upflow over 3 L of ¼-in. tab alumina balls (Harshaw Chemical Company) at a nominal pump rate of 1 L/hr. Pressure was maintained at 2000 psig with once-through hydrogen during the run. Temperature at the bed inlet was varied between 600° and 800°F. Data from the run are summarized in Table II.

Other than a slight reduction in sulfur content the filtrate appeared to be thermally stable. Heptane insolubles showed little consistent change. This would be expected since these conditions are very similar to the solvent-refining process environment.

The residual fraction in the filtrate, which is undistilled 850°F+ SRC, was later found to be thermally unstable. This was determined from examination of the products before and after distillation using

Table I.　Tacoma Solvent-Refined Products[a]

Designation	Total Filtrate	850°F+ SRC Residue
Wt % of total filtrate	100	24.3
Ultimate analysis (wt %)		
C	86.9	87.8
H	7.6	5.6
N	1.0	2.3
S	0.55	0.66
ASH	0.033	0.19
Density (20/4°C) (g/mL)	1.068	1.142
Softening point (°F)	—	352
Pour point (°F)	—	440
Distillation, vapor (°F)		
IBP	—	563
5	303	883
9	—	915
10	409	—
30	467	—
50	561	—
70	668	—
80	728	—
Benzene insolubles (wt %)	9.4	48
Heptane insolubles (wt %)	16.4	87

[a] Kentucky Colonial coal run June 21, 1976.

Table II. Filtrate Processed Over TAB Alumina[a]

Hours on-stream	Feed	9	22	30	38
Temperature (°F)		600	650	800	800
Chemical analysis (wt %)					
C	86	87	87	—	88
H	7.7	7.5	7.7	—	7.5
N	1.1	1.1	1.1	—	1.1
S	0.55	0.50	0.46	0.45	0.39
Sulfur removal (wt %)	—	9.1	16	18	29
Nitrogen removal (wt %)	—	0	0	—	0
Heptane insolubles (wt %)	16.6	15.6	14.0	16.2	17.9

[a] Pressure = 2000 psig; H_2 Feed Rate = 450 std L/hr; Space Velocity = 0.3–0.4 hr^{-1}.

heptane- and benzene-solubility tests. In order to reach the 750°F cut point used in these filtrate distillations the pot temperature often went as high as 650°F. The solubility properties of the residual product taken from the distillation flask were different from undistilled product (Table III). Since all insolubles were in the 750°F residue, heptane insolubles had increased by 40% for the distillation bottoms material over values for undistilled product; benzene insolubles increased by 18%.

However, the heptane insolubility of 850°F+ SRC residue from Tacoma (see Table I) showed little change when subjected to the same treatment as the filtrate from Tacoma. The 850°F+ SRC residue received from Tacoma had 81% heptane insolubles while the 850°F+ residue from the distillation of the reconstituted SRC filtrate, which is described below, had 87% heptane insolubles. This was a 6% increase in heptane insolubles in Tacoma residual SRC after thermal treatment compared with a 40% increase for the Tacoma filtrate heptane insolubles after the same treatment.

Apparently the increase in heptane insolubles resulted from thermal repolymerization during distillation. The filtrate readily underwent this reaction but Tacoma 850°F+ SRC residue was unreactive because it already had been exposed to high temperature when isolated by distilla-

Table III. Heptane and Benzene Insolubles of Distilled and Undistilled Tacoma Samples

	Heptane	Benzene
Tacoma Filtrate		
Undistilled (wt %)	16.4	9.4
Distilled (wt %)	23.0	11.1
Change (%)	40	18
Tacoma 850°F+ SRC Residue		
Undistilled (wt %)	81	46
Distilled (wt %)	87	48

tion at an 850°F cut point. Further heating during subsequent distillations failed to further polymerize the material. But, based on our results presented below, the further hydrorefining of this 850°F+ SRC residue rendered it thermally unstable.

Filtrate Processing over Co–Mo Catalyst. Tacoma filtrate was processed upflow over a fixed bed of Co–Mo supported on silica stabilized ⅛ × ⅛-in. extruded alumina catalyst obtained from Harshaw Chemical Company (Co–Mo 0402-T). Analyses of the fresh catalysts used in this study are shown in Table IV. This is the preferred catalyst for the Synthoil process (1). The process conditions used for this catalyst were within the following ranges: temperature, 650–800°F; volume space velocity, 0.3–1.6 hr^{-1}; hydrogen feed rate, 1400–7400 scf/bbl; pressure, 1500–2000 psig. Gaseous products were not collected, measured, or analyzed during the filtrate runs. Data calculations are based entirely on recovered liquid products. Recoveries ranged from 97 to 107 wt % with an average recovery of 99.8 wt %. Data from filtrate processing are presented in Table V where a 750°F endpoint was used. In Table VI, data for an 850°F cut point are presented.

Conversion of SRC. Increased distillate product yield recently has been the primary aim for further processing of SRC, although sulfur removal has gained renewed importance. Conversions were determined primarily from distillation data using a short path still under a nominal 0.1 mm Hg pressure. An IBP to 420°F light cut, a 420°F to higher cut point distillate recycle solvent cut (either 750°F or 850°F), and the bottoms residual fractions, along with condensed water, were determined.

Table IV. Commercial Hydrorefining Catalysts[a]

Source	American Cyanamid	Harshaw	Ketjenfine
Designation	HDS 3A	Co–Mo 0402	HC
Ni	2.9	—	4.6
MoO$_3$	12.8	15.0	—
CoO	—	3.0	—
W	—	—	16.0
Al$_2$O$_3$	—	—	18.0[b]
Na$_2$O	—	—	.01
SO$_4$	—	—	1.0
Fe	—	—	.03
Pore volume (mL/g)	.4	.34	.66
Surface area (m^2/g)	180	190	187
Average pore, diameter Å	40–100	—	90

[a] Compositions are in weight percent.
[b] Remainder silica.

Table V. Filtrate Processed over Co–Mo Catalyst

	Feed	19	76	83	101	134
Hours on-stream	Feed	19	76	83	101	134
Temperature (°F)		650	750	750	750	750
Pressure (psig)		2000	2000	2000	2000	2000
LHSV (hr⁻¹)		1.0	0.7	1.0	0.3	1.6
H_2 rate (scf/bbl)		2100	3100	2400	7400	1500
Recovery (wt %)		97.1	100.7	94.6	107.4	100.4
H_2 consumption (scf/bbl)		150	570	650	1080	720
Chemical analysis (wt %) [a]						
C	88.8	87.8	87.9	88.0	86.8	87.4
H	7.5	8.1	9.1	9.0	9.7	8.6
N	1.10	1.04	0.89	0.94	0.80	0.98
S	0.53	0.32	0.08	0.11	0.07	0.13
O	4.0	2.8	1.4	2.88	2.7	
Product distribution (wt %)						
IBP–420°F		7.7	6.3	8.1	5.9	4.8
420–750°F	71.6	60.5	63.7	62.1	74.2	64.5
750°F+	28.4	31.4	27.5	26.9	18.5	26.3
water		0.3	2.1	2.2	1.3	4.3
SRC converted (wt %) [b]		−8.0	3.7	5.8	35.4	8.0
Solvent converted (wt %) [c]		10.8	8.8	11.3	8.3	6.8

[a] Upon distillation of these total products in this run, copious amounts of ammonia and water were observed upon distillation. Subsequent runs were made at higher separator temperatures to eliminate this problem.

[b] SRC converted is the weight percent loss of 750°F+ feed based on recovered product.

[c] Solvent converted is the weight percent loss of 420°–750°F feed based on that recovered.

Table VI. Results on Tacoma Filtrate Using 850°F Cut Point for Co–Mo

	Feed	76	83	101
Hours on-stream	Feed	76	83	101
Temperature (°F)	—	750	750	750
Pressure (psig)	—	2000	2000	2000
LHSV (hr⁻¹)	—	0.7	1.0	0.3
Recovery (wt %)	—	100.7	94.6	107.4
Product distribution (wt %)				
IBP–420°F [a]	—	11.5	8.6	12.3
420°–850°F	77.2	67.9	71.5	73.9
850+°F	24.3	17.9	18.1	12.9
Water	—	2.7	1.9	0.9
850°F+ SRC converted (wt %) [b]	—	26	26	47
Solvent converted (wt %) [c]	—	10.4	5.7	2.5

[a] The difference in IBP to 420°F weight percents shown here and in Table V is a result of experimental differences. A short distillation head was used in both distillations, thereby resulting in the experimental inaccuracies for this cut.

[b] SRC converted refers to the weight percent loss of 850°F+ feed based on recovered product.

[c] Solvent converted is the weight percent loss of 420°–850°F feed based on recovered product.

The relative change in SRC conversion vs. hydrogen consumption is seen in Figure 3 for both 750° and 850°F cut point data. As hydrogen consumption increases, SRC conversion also increases. The intercept for the 750°F cut point data suggests that 0.8 wt % hydrogen addition to the feed is necessary before any conversion of the residue is observed. This type of result also has been observed in our studies on synthetic filtrate.

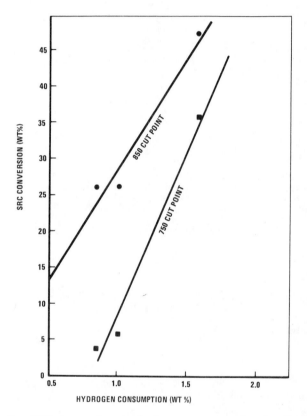

Figure 3. SRC conversion vs. hydrogen consumption for Co–Mo: (■), 750°F cut point; (●), 850°F cut point.

HETEROATOM ELIMINATION. Sulfur elimination from hydrorefined filtrate varied from 42 wt % at 650°F to 98 wt % under the most severe conditions at 750°F. The data presented in Figure 4 shows a very rapid initial decrease in sulfur content probably related to an initially rapid catalytic conversion of sulfur compounds particularly susceptible to catalytic reaction but thermally stable. Such compounds could be di-

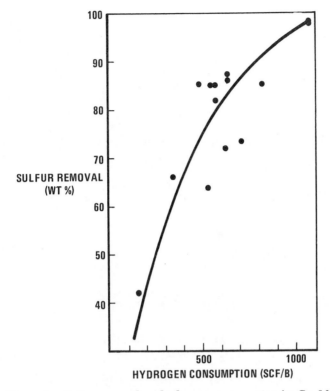

Figure 4. Sulfur removal vs. hydrogen consumption for Co–Mo

benzothiophene or partially substituted dibenzothiophenes. Less reactive sulfur atoms are ones which are unable to approach the active catalyst site because of steric considerations (2).

The data for oxygen removal vs. hydrogen consumption is shown in Figure 5. As hydrogen consumption increases oxygen removal increases. The line drawn in this figure represents a hydrogen:oxygen atom ratio of 9:1 which represents the number of hydrogen atoms necessary to remove one oxygen atom.

In the liquefaction process the hydrogen consumption in eliminating oxygen from coal is an important factor. In the initial coal liquefaction reaction as shown in the data of Whitehurst et al. (3) two hydrogen atoms are consumed for each oxygen atom removed. In a later stage, consumption increased four to six hydrogen atoms per oxygen atom removed. For the hydrorefining of filtrate feed we might expect ratios of hydrogen to oxygen on the order of 4:1 or greater since these filtrates already have been through the early stage of oxygen removal. Therefore, the 9:1 ratio we observed is entirely within our expectation.

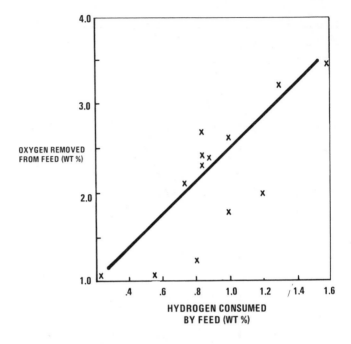

Figure 5.   Oxygen removal vs. hydrogen consumption for Co–Mo (oxygen in feed 4%)

The sensitivity of conversion to final cut point is observed by the differences for three samples.

| | SRC Conversion (wt %) | |
Sample	750°F+	850°F+
CP3–23	4	26
—25	6	26
—27	35	47

An 850°F cut point was used at the Tacoma plant when this filtrate was generated.

Comparing the 750°F+ and 850°F+ residual feed conversion above indicates sizable conversion for the 850°F+ SRC (26–47%) and considerably smaller conversion for 750°F+ SRC feed (4–35%). This leads to the conclusion that the major product from SRC conversion was material boiling between 750° and 850°F. For samples taken at 76 and 83 hr on-stream, approximately 20% of the feed SRC went to 750°–850°F distillate. Only about 5% went to 750°F distillate since further conversion to 750°F material was much slower. For the sample taken at 101 hr on-stream, the residual SRC conversion for 850°F and 750°F cut points, respectively, dropped from 47% to 35%, suggesting that intermediates in this 750°–850°F range are less reactive but can be converted under the more severe conditions.

Nitrogen removal from total feed as calculated from elemental analyses of distillate and recovered bottoms varied between 23 and 62 wt %. Trapped water and ammonia were observed when the recovered products were distilled, which accounts for the disparity with data in Table V. In subsequent runs this problem was eliminated by using higher separator temperatures. These two values represented the mildest and the most severe conditions used during this run. These results suggest a definite correlation between severity and elimination. Other data, however, fail to show a smooth correlation.

Processing Reconstituted SRC Filtrate. A synthetic blend of 850°F+ residual SRC in a coal-derived creosote oil was prepared and used in catalyst evaluation studies when SRC filtrate became unavailable. The synthetic blend was a 52 wt % mixture of residual Tacoma 850°F+ SRC dissolved in Koppers Tar creosote oil.

Creosote oil was chosen as substitute solvent because of its coal source and its similarity in ring composition to SRC distillate. The hydrogen contents of these oils are lower than for conventional recycle solvents. Koppers Tar has a 6.1% hydrogen content whereas the SRC distillate from Tacoma used in this study has an 8.2% hydrogen content.

When used as solvent in the SRC process the hydrogen donor capacity of creosote oils was improved by hydrotreating. Hydrotreating increased the hydrogen content up to 1 wt % to bring them more in line with typical recycle solvents (4). However, in this program we proceeded without prehydrotreating of the solvent because of the press for time. Conceptually, the experimental data would better reflect residual SRC conversion and selectivity if hydrogen consumption for upgrading solvent were minimized.

Koppers Tar creosote oil is comprised primarily of two-, three-, and four-ring polynuclear aromatic compounds. Only 4% of this solvent, which contains a sizable amount of fluoranthene (bp 723°F), boils over 850°F. Typical hydrotreated 420°–850°F distillate product from this feed material has a multitude of additional components in much smaller concentrations as determined by gas chromatography (GC). The major components of the feed contribute less to the total concentration in product than in feed. Therefore, product distillates are less pure than feed distillates and less likely to solidify than feed. Indeed, Koppers Tar is itself a solid–liquid mixture which upon observation suggests handling problems. In anticipation of pumpability problems for mixtures of SRC in Koppers Tar, a viscosity study was made on several different concen-

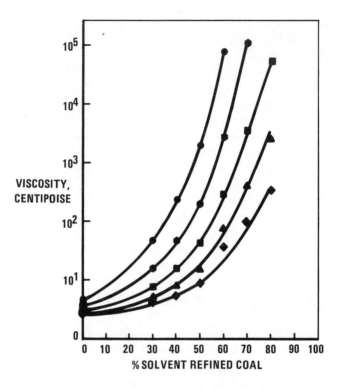

Figure 6. Viscosity of reconstituted SRC filtrate vs. composition: (●), 200°F; (◯), 250°F; (■), 300°F; (▲), 350°F; and (◆), 400°F.

trations of SRC in Koppers Tar at different temperatures (Figure 6). A feed mixture was used which contained 52% 850°F+ SRC and had a viscosity below 8 cP at 400°F.

FEED PREPARATION. The feed was prepared by adding the SRC to Koppers Tar heated to 300°F in a 60-gal tank under a nitrogen blanket. Solvent was added to the tank under nitrogen and heated to 300°F. SRC was added initially in large (30–50 lb) batches and later in smaller (10-lb) increments as the concentration increased. When the concentration was close to 50%, the tank was heated to 400°F. The solution was mixed for a day before being used. Homogeneity was checked by examining samples removed from the tank. A 400°F temperature was maintained in the mix tank, feed tank, transfer lines to the reactor, and the pump head during the run. Transfer lines and vessels downstream from the reactor were kept at 300°F. From distillation results on this feed we determined that the concentration of 800°F+ material in the charge material was 57.4 wt %; the 850°F+ material was 52 wt %.

CATALYST EVALUATION. Commercial nickel–molybdenum (Ni–Mo) and nickel–tungsten (Ni–W) catalysts were evaluated with this feedstock. The Ni–Mo catalyst was HDS-3A from American Cyanamid and the Ni–W catalyst was Ketjenfine HC-5 from Armak Company. Both were extrudate types supported on alumina and silica–alumina, respectively. The run conditions for the Ni–W evaluation run are shown in Table VII for selected samples. Pressure, liquid feed rate, and hydrogen feed rates were held as nearly constant as possible; only the temperature was changed.

The conversion of 850°F+ residual SRC to lighter product was determined for several balance periods in which recoveries varied between 98.5 and 100.4 wt %. The distillations were run at a slower rate than for the filtrate studies in order to sharpen the 850°F cut point for better data reproducibility. Product distributions are shown in Figure 7.

A net solvent make from 850°F+ SRC conversion was observed at all conditions with only slight temperature sensitivity. The yield of 420°–850°F solvent ranged from 48.6 to 53.4 wt % over a 120°F temperature range. Light distillate make increased slightly with temperature. The yield of hydrocarbon gaseous product varied between 0.4 and 4.3 wt % of feed. The gas based on converted 850°F+ SRC varied between 7 and 38 wt % with increasing temperature. The compositions of the gaseous hydrocarbon products are shown in Table VIII.

The order of concentration follows the descending order: methane > ethane > propane > butane > C_5^+. Olefinic materials are almost non-existent in these products. The yields of H_2S, H_2O, and NH_3 increase as temperature increases going from 1.0 to 3.6 wt % of feed. At their highest yields, 86% of the sulfur, 70% of the oxygen, and 38% of the nitrogen were removed from the liquid feed.

Table VII.   Data Summary Sheet for Ni–W Catalyst Evaluation

Hours on-stream[a]	24	54	66	102	126
Balance period (hr)	5.5	5.5	5.5	5.75	5.5
Pressure (psig)	2000	2000	2000	2000	2000
Hydrogen rate (SLPH)	313	333	320	330	367
LHSV (hr⁻¹)	0.67	0.60	0.59	0.53	0.61
Liquid feed (g/hr)	689	646	629	564	650
Temperatures (°F)					
Bed entrance	712	763	763	790	813
Bed middle	736	794	793	819	844
Bed exit	722	772	772	797	821
Main receiver product (g/hr)	681	612	602	524	593
Product gas rate (SLPH)	221	193	192	182	213
Recovery (wt %)	100.4	98.5	99.6	99.3	99.9

[a] End of balance.

Table VIII. Hydrocarbon Gas Distribution for Ni–W (Wt %)

Temperature (°F)	726	736	784	794	793	820	844
Methane	36	40	39	34	34	30	33
Ethane	36	30	32	28	27	24	25
Ethylene	0	0	0	0	0	0	0
Propane	18	17	18	20	20	22	20
Proplyene	0	0	0	0	0	0	0.3
Butanes	5	10	8	13	13	17	14
Butylenes	0	0	0	0	0	0	0
C_5^+	5	3	3	6	6	7	5

Conversion of 850°F+ SRC feed is presented in Figure 8 as a function of temperature. As temperature increases conversion of 850°F+ SRC also increases.

Hydrogen consumption values were calculated by two independent methods. Hydrogen balance values were obtained by comparing hydrogen concentrations at the inlet and outlet of the reactor. A totally independent

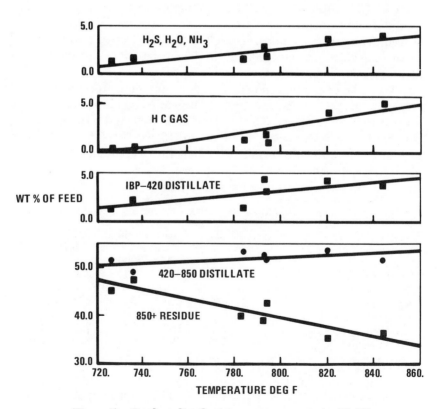

Figure 7.   Product distribution vs. temperature for Ni–W

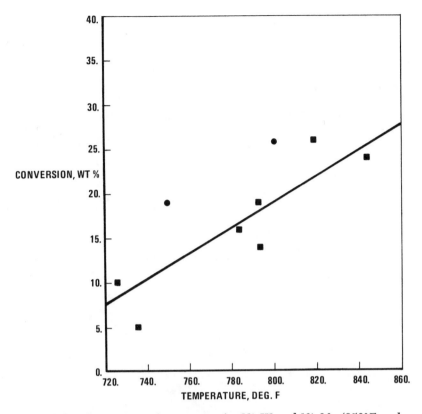

Figure 8. Comparison of conversion for Ni–W and Ni–Mo (850°F endpoint): (■), SRC conversion for Ni–W; (●), SRC conversion for Ni–Mo.

elemental hydrogen consumption was calculated from the increased combined hydrogen in the product vs. the feed. These comparisons for the Ni–W case presented in Table IX are in good agreement. As the plot of data in Figure 9 shows, hydrogen consumption increases as conversion of 850°F+ SRC increases.

Because we find that a linear relationship exists between temperature and hydrogen consumption on the one hand and between temperature and SRC conversion on the other, a linear relation should exist between hydrogen consumption and SRC conversion. Indeed, the plot of data in Figure 10 shows such a relationship.

As hydrogen consumption increased, the conversion of 850°F+ SRC to 850°F distillate product also increased. Significantly, a certain threshold of hydrogen had to be consumed before any SRC was converted to distillate. The intercept in Figure 10 shows that the threshold level is about 0.5 wt % hydrogen consumption.

	Table IX.	Hydrogen
Temperature (°F)	726	736
H_2 feed rate (scf/bbl)	2800	2600
Consumption by gas balance (in–out (scf/bbl)	850	850
Consumption: by elemental balance (scf/bbl)	650	800
Consumption on feed (wt %)	0.95	1.20

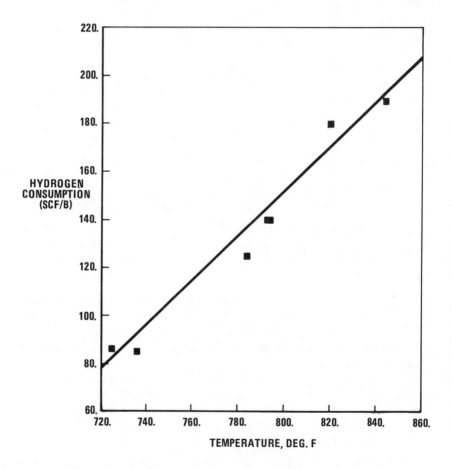

Figure 9.   Hydrogen consumption vs. temperature for Ni–W

Consumption for Ni–W

784	794	793	820	844
3000	3100	3000	3500	3400
1250	1400	1400	1800	1900
1050	1200	1350	1700	1600
1.59	1.78	2.00	2.56	2.44

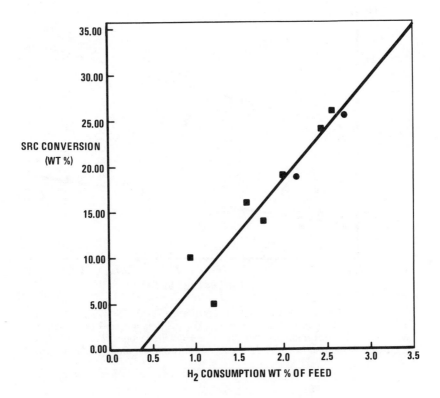

Figure 10. 850°F + SRC conversion vs. total hydrogen consumption for Ni–W and Ni–Mo: (■), conversion over Ni–W; (●), conversion over Ni–Mo.

The heteroatom elimination from total feed is plotted vs. temperature in Figure 11. The data show an increase in sulfur, oxygen, and nitrogen removal as the temperature increases. The scatter in the oxygen data is considerably greater than for either nitrogen or sulfur.

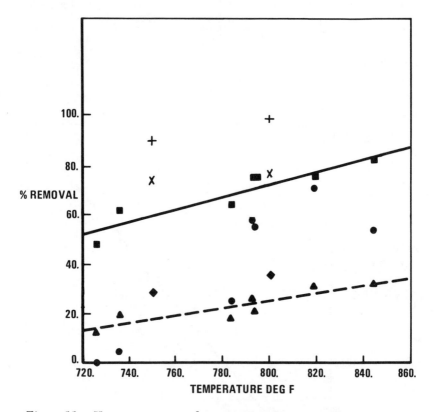

Figure 11.   Heteroatom removal vs. temperature for Ni–W and Ni–Mo:
(■, ——), sulfur removal over Ni–W; (●), oxygen removal over Ni–W;
(▲, – – –), nitrogen removal over Ni–W; (+), sulfur removal over Ni–Mo;
(×), oxygen removal over Ni–Mo; and (♦), nitrogen removal over Ni–Mo.

On a relative gram-atom removal basis the number of oxygen atoms removed at the higher temperatures is larger than for nitrogen which is greater than for sulfur (Table X). Far more hydrogen is consumed in removing oxygen and nitrogen than in removing the small amount of sulfur present.

Table X. Heteroatom Elimination from Total Product for Ni–W

Temperature (°F)	726	736	784	794	793	820	844
Relative gram–atoms removed							
Sulfur	1.0	1.0	1.0	1.0	1.0	1.0	1.0
Nitrogen	1.6	1.8	1.7	1.7	2.1	2.4	2.3
Oxygen	—	0.6	3.5	6.6	7.0	8.4	5.9

Comparison with Ni–Mo Catalyst. Ni–Mo catalyst was evaluated in a run similar to that of Ni–W, but at only two temperatures, 750° and 800°F. Superimposing the 850°F+ cut point conversion values of Ni–Mo onto the data for the Ni–W catalyst shows a definite advantage of Ni–Mo over Ni–W at both temperatures (Figure 8). At 750°F we have an activity advantage of 10% SRC conversion; at 800°F the difference is a 13% SRC advantage. From a temperature standpoint the Ni–Mo catalyst used in this study has about a 50°F advantage over the Ni–W catalyst used here. A generic advantage of Ni–Mo over Ni–W cannot be drawn from the data obtained in this study.

A comparison of 850°F+ SRC conversion vs. hydrogen consumption is shown in Figure 10. The efficiency of hydrogen utilization is extremely important in this process where hydrogen costs will be so great. These data show no significant difference in conversion vs. hydrogen consumption.

A comparison of the removal of sulfur, nitrogen, and oxygen from liquid feed is shown in Figure 11. Removal levels for each of the three heteroatoms is greater for Ni–Mo than for Ni–W.

A similar comparison of heteroatom removal vs. hydrogen consumption is presented in Figure 12. The results show that for Ni–Mo at the same hydrogen consumption level, sulfur, oxygen, and nitrogen removal are all greater than for Ni–W. This occurs even though the temperature of reaction at these hydrogen consumption levels are quite different. At even the most severe condition for the Ni–W run, the level of each heteroatom removal is less than for Ni–Mo run at 750°F. Clearly, Ni–Mo is a far better catalyst for removal of heteroatoms than Ni–W.

Converted product selectivity on a carbon basis is compared for Ni–W and Ni–Mo in Figure 13 for 850°F+ SRC feed. The yield of hydrocarbon gases from Ni–W is much greater than for Ni–Mo. Part of the reason for the difference may be the higher temperature necessary to attain the same level of conversion for Ni–W than for Ni–Mo. In any case the much higher selectivity of Ni–Mo for liquid distillate at the same level of hydrogen consumption to reach a particular conversion makes the Ni–Mo the preferred catalyst over Ni–W.

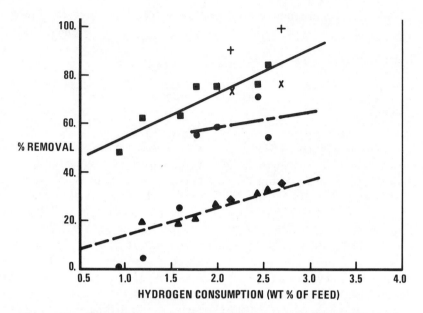

Figure 12.   Heteroatom removal vs. hydrogen consumption: (■,——), *sulfur removal over Ni–W;* (●, – – –), *oxygen removal over Ni–W;* (▲, – – –), *nitrogen removal over Ni–W;* (+), *sulfur removal over Ni–Mo;* (×), *oxygen removal over Ni–Mo; and* (♦), *nitrogen removal over Ni–Mo.*

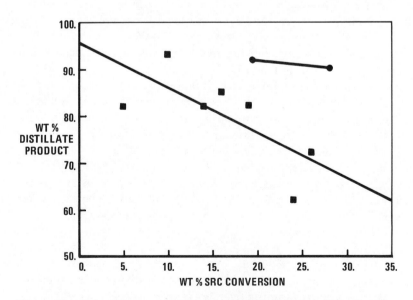

Figure 13.   Selectivity vs. conversion for 850°F cut point: (■), *selectivity for Ni–W;* (●), *selectivity for Ni–Mo.*

THERMAL INSTABILITY OF HYDROREFINED PRODUCTS. Thermal instability of hydrorefined product from the reconstituted feed was observed
for both Ni–W and Ni–Mo. Comparing the conversion of heptane insolubles for undistilled and distilled materials showed evidence of
considerable polymerization for the distilled material. Heptane insolubles
conversion for undistilled product from Ni–W was higher than for distilled
product as seen in Figure 14. Distilling the product from the 725°F
balance reduced the heptane-insoluble conversion from 30% to 16%.
For Ni–Mo the reduction for the heptane insolubles was on the same order
of magnitude, approximately a 10% decrease in the heptane-insolubles
conversion.

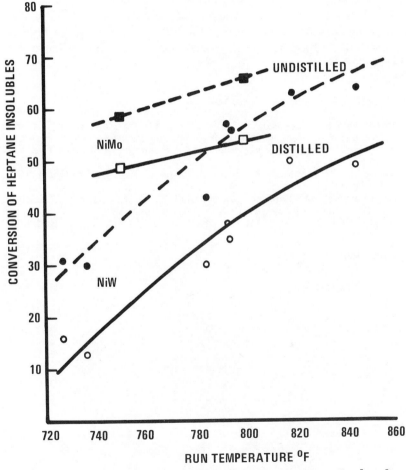

*Figure 14. Heptane-insoluble conversion vs. temperature. Total undis-
tilled sample: (■), Ni–Mo; (●), Ni–W. Bottoms from distillation: (□),
Ni–Mo; (○), Ni–W.*

Benzene-insolubles conversion was essentially unchanged upon distillation as shown by the Ni–Mo data below.

| | Conversion of Benzene Insolubles (wt %) ||
Ni–Mo Run at	Undistilled	Distilled
750°F	75	76
800°F	76	75

Ni–W data are unavailable since these measurements were cancelled because of the newly published restrictions on benzene exposure levels.

The thermal instability of hydrorefined SRC creates another parameter which requires careful consideration. If, for example, topping operations reverse some of the depolymerization occurring in the SRC process, this equates to a hydrogen penalty. Additional work needs to be done to characterize the asphaltenes generated thermally vs. the starting asphaltenes from coal.

Literature Cited

1. Sayeed Sayeed; Mazzocco, N. J.; Weintraub, M; Yavorsky, P. M. "Synthoil Process for Converting Coal to Nonpolluting Fuel Oil," paper presented at the 4th Synthetic Fuels from Coal Conference, Stillwater, OK, May 1974.
2. Givens, E. N.; Venuto, P. B. *Prepr., Am. Chem. Soc. Div. Pet. Chem.* **1970,** *15*(4), A183.
3. Whitehurst, D. D.; Farcasiu, M.; Mitchell, T. O. Energy Power Research Institute Annual Report, EPRI RP410-1, February 1976.
4. Wright, C. H.; Pastor, G. R.; Perrussel, R. E., ERDA Report FE496-T4, September 1975.

RECEIVED June 28, 1978.

Chemicals from Coal-Derived Synthetic Crude Oils

ROBERT D. CALDWELL and STEPHEN M. EYERMAN

The Dow Chemical Company, Midland, MI 48640

Four different coal-derived oils were evaluated as potential petrochemical feedstocks. The oils were distilled and mid-distillate hydrocracking was done on alumina-supported cobalt–molybdenum catalyst. Temperatures from 930° to 960°F were necessary to achieve 60% conversion at 2500 psig. Hydrotreating the naphthas to create acceptable reformer feedstocks proved most difficult, requiring two passes on most straight-run naphthas to reduce sulfur and nitrogen to less than 5 ppm. Reforming, done on a platinum catalyst, yielded 85–92% aromatics in the C_{6+} reformate at 968°F and 500 psig.

When one considers the status of our fossil fuel resources and the factors shaping their future use patterns, it becomes clear that in spite of the problems associated with coal, its availability could be an important factor in reducing resource shortages. It is apparent that prompt, optimum usage of coal should be developed. It is not a question of chemical or energy needs, but a necessity that both needs be met. In the crisis atmosphere related to energy, the dependence of chemical requirements on fossil fuels is commonly overlooked. It was with this in mind that The Dow Chemical Company proposed, and was subsequently granted, a contract by The United States Energy Research and Development Administration (ERDA) to evaluate coal-derived liquid products as petrochemical feedstocks.

That coal-derived liquids consist largely of aromatic and cyclic structures suggests that, with appropriate hydroprocessing, high yields of benzene, toluene, and xylenes (BTX's) should be obtained. If the by-products of that BTX production could be directed toward linear paraffins, useful olefins also might be produced to supplement the benzene value. This study was limited to using standard hydroprocessing operations to ascertain a preliminary value of the coal-derived oil as feedstock. No attempts were to be made to recover tar acids and/or bases, and only readily available commercial catalysts were used.

The objective of the study was to evaluate the suitability of four different coal-derived liquids as petrochemical feedstocks. The four liquids received are identified further as follows:

COED (Char Oil Energy Development, FMC Corp.)—hydrotreated pyrolysis oil, 16.6 wt % on dry Western Kentucky coal, and < 10 cP, 900°F endpoint.

H–Coal (Hydrocarbon Research, Inc.)—catalytic hydrogenation of coal in an ebulliated bed reactor (Syncrude mode), 80.3 wt % on dry Illinois #6 coal, 1 part distillate of < 10 cP and 2 parts high mp solid containing 32 wt % solids.

SRC II (Solvent-Refined Coal (SRC) II, Pittsburg and Midway Coal Mining Co.)—hydroliquefaction of coal with slurry recycle, 25.3 wt % on dry Western Kentucky coal, < 10 cP 850°F endpoint.

Synthoil (U.S. Bureau of Mines—Bruceton)—catalytic hydroliquefaction of coal in a fixed-bed reactor, 59.3 wt % on dry West Virginia coal 30SSU @ 180°F 50% 850°F+.

Experimental

The following is the processing sequence used for all of the samples. (1) Distillation into nominally three cuts: IBP–350°F straight run naphtha; 350°–650°F mid-distillate; 650°–850°F gas oil (where practicable). (2) Hydrocracking of the mid-distillate and gas oil to produce naphtha and LPG's; (3) hydrotreating of the straight run naphtha and hydrocrackate naphtha to remove heteroatoms; and (4) reforming of the hydrotreated naphthas to maximize aromatics.

Distillations were carried out in a 72-L Podbielniak Fractioneer A batch still which had been modified to run unattended. Initial fractionation results are contained in Table I.

Hydroprocessing studies were carried out in continuous fixed-bed tubular reactors containing 50–200 cm^3 of catalyst. All reactors were operated in downflow mode. To facilitate mass balance data acquisition,

Table I. Composition of "As Received" Samples (Wt %)

	COED	H–Coal	SRC II	Synthoil
Naphtha (IBP–350°F)	22.4	19.8	22.0	0
Mid-distillate (350°–650°F)	46.2	16.5	63.0	34
Gas oil (650°–850°F)	27.8	7.5	15.0	16[a]
Residue	3.6	56.2[b]	0	50

[a] Data derived from simulated distillation, gas oil not recovered.
[b] 25% Ash, 13% unreacted coal, 62% benzene soluble.

liquid feed tanks were set upon scales while product receivers were weighed periodically. Hydrogen and vent gas rates were determined with calibrated integral orifice flowmeters. A square root integrator provided a time-weighted average vent gas rate.

On-line analysis of the hydrocarbons in the vent gas was done on a 20-ft Porapak Q column with an internal standard. In many cases, the internal standard also allowed calculation of hydrogen by difference. A Beckman 3AM3 gas density balance provided a check on the vent gas analysis and the necessary parameter for calculating mass flow rates from orifice pressure drop data. Mass balance closure was typically better than ± 2%.

The hydroprocessing reactors were allowed to run unattended for the afternoon and evening shifts. Therefore, 24-hr runs were the norm and, in the event of analytical lag, some runs were extended for a period of days.

The following inspections on liquids were performed as necessary.

(1) Carbon–hydrogen: use of the Perkin Elmer Model 240 CHN analyzer required encapsulation of the volatile liquids in quartz vials. A Model 1200 Chemical Data Systems elemental analyzer was used also.

(2) Sulfur: the Dohrmann microcoulometric reactor was used in the oxidative mode.

(3) Nitrogen: a Dohrmann microcoulometric reactor being used in reductive mode was replaced with an Antek Model 771 pyroreactor which uses a chemiluminescent nitrogen detector.

(4) Oxygen: a Kaman Model A711 neutron generator provides 14.3×10^6 eV as activation energy for neutron activation analysis.

(5) Water: the Aquatest I does a coulometric Karl Fischer titration of water dissolved in oils.

(6) Naphtha componential: analysis of C_3–C_9 hydrocarbons was done on a 200-ft Squalane capillary column with a flame ionization detector (FID).

(7) Simulated distillations: the boiling range was obtained on an 8-ft bonded methyl silicone column using a FID (*see* ASTM D-2887). A simple internal standard variation was used for nondistillates.

(8) Mercury and Gallium: a General Electric TRIGA reactor provides slow neutrons for the neutron activation analysis of these metals at the 10–50-ppb detection level.

Data reduction to usable form was done via a series of computer programs. Programs were written for vent gas analysis, naphtha componentials, simulated distillation, elemental balances, and overall mass balances with C_1–C_9 componential yields.

Results and Discussion

Hydrocracking. Conventionally, the term hydrocracking is used to describe a process wherein hydrogen and a distillate hydrocarbon are passed over a bifunctional catalyst which provides both a hydrogenation function and an acidic site cracking function. Evidence of acidic site titration by nitrogen and/or oxygen is not uncommon. Expecting such deactivation of conventional catalysts by the high levels of oxygen and nitrogen in coal-derived mid-distillates and gas oils led to the choice of cobalt–molybdenum (Co–Mo) on alumina as the catalyst for hydrocracking. The catalyst would provide a hydrogenation function, while the cracking would be largely thermal. This type of operation would yield much more liquefied petroleum gas (LPG) and methane than conventional hydrocracking, but LPG's are not unattractive as olefin feedstocks and SNG is becoming increasingly valuable.

The catalyst used in the mid-distillate and gas–oil hydrocracking was Harshaw 400T. The COED, H-Coal, and Synthoil samples were run over 1/8-in. extrudates; the SRC II sample was run on 1/16-in. extrudates. The catalyst was presulfided in situ with 1% CS_2 in a light naphtha.

Inspections on the four mid-distillates are presented in Table II. The 560°F endpoint for the H–Coal mid-distillate is caused by the fact that the 560°F+ material was delivered as process residue containing the ash and unreacted coal. The residue material was very difficult to distill in a batch system. Scheduling pressures forced the H–Coal hydrocracking runs to be done on the available material.

Owing to the limited availability of the feedstocks, it was not possible to do parametric studies and then operate at a single set of conditions to provide a hydrocrackate naphtha. All of the hydrocrackate material from a given feedstock was combined, the naphtha distilled, and resultant unconverted mid-distillate subjected to second-pass hydrocracking.

A typical data set of the over 70 accumulated is presented in Table III. The hydrogen yield (consumption) was calculated for the mass balance assuming that the unanalyzed fraction of the vent gas was hydrogen. Since water vapor, hydrogen sulfide, and ammonia are not determined analytically in the vent gas, the hydrogen yield as calculated from vent gas analysis is in error. The elemental balance consists of an analytical carbon–hydrogen balance, with ammonia, hydrogen sulfide,

Table II. Mid-Distillate Inspections (Wt %)

	COED	H–Coal	SRC II	Synthoil
Carbon	88.3	88.1	87.1	86.3
Hydrogen	11.2	11.2	8.4	9.3
Oxygen	0.34	0.49	3.9	3.3
Nitrogen	0.16	0.044	0.34	0.30
Sulfur	0.005	0.17	0.30	0.74
H/C atomic ratio	1.51	1.51	1.16	1.28
TBP range (°F)	350–690	350–560	360–680	360–680
Hg	< 20 ppb	< 60 ppb	< 50 ppb	210 ppb
Ga	< 10 ppb	< 1 ppb	< 15 ppb	120 ppb

Table III.   Hydroprocessing Data

Process: Hydrocracking Mid-Distillate
Catalyst: Harshaw HT-400-1/8-E
Feed: Synthoil Straight-Run Mid-Distillate

Reactor Conditions

pressure (psig)	2510
LHSV (vol/vol–hr)	0.88
material balance (%)	99.45
hydrogen ratio (scf/bbl)	20460
oil feed rate (g/hr)	87.43
hydrogen feed rate (g/hr)	28.81
temperature (°F)	947
catalyst volume (mL)	100
force balanced on	—
oil product rate (g/hr)	50.2
tail gas rate (g/hr)	63.03
water product (g/hr)	2.37

Conversions weight percent
450°F +   80.5
550°F +   85.8

Yield: G/100G oil feed

hydrogen	−3.54[a]	−4.39[b]
methane	4.04	
ethane	5.82	
propane	6.65	
water	2.71	

	Paraffin		Naphthene			
	Normal	*Iso*	*Cyclo C_5*	*Cyclo C_6*	*Aromatic*	*Total*
C_4	5.19	.81				5.99
C_5	1.37	.78	1.00			3.15
C_6	1.15	.46	3.38	2.91	3.80	11.70
C_7	.74	1.54	.81	3.74	6.37	13.20
C_8	.44	.19	.84	1.36	3.31	6.14
C_9	.19	.20	.08	.25	2.09	2.82
Total	9.08	3.97	6.11	8.27	15.57	43.00

Unidentified
C_4–C_9 0.0
C_{10}+   40.55

Hydrogen yield (scf/bbl)
−2190[a]   −2725[b]

Elemental material balance (wt)

Table III. Continued

	Carbon	Hydrogen	Nitrogen	Oxygen	Sulfur	Total
Feed						
oil	86.3	9.3	.28	3.3	.74	100.00
hydrogen		32.95				32.95
total	86.3	42.25	.28	3.3	.74	132.95
Products						
liquid	50.93	6.14	.105	.208	.022	57.41
HC gas	35.57	7.07				42.65
hydrogen		29.42				29.42
NH_3		.04	.18			.22
H_2O		.39		3.09		3.48
H_2S		.05			.72	.77
total	86.50	43.11	.28	3.3	.74	133.95

[a] From hydrogen mass balance.
[b] From elemental balance.

and water calculated on the basis of the change in heteroatom concentration in the liquids. The conversion data were correlated in terms of the following semi-empirical expression: $\ln(1-C) = -Ae^{-E/RT} P_{H_2}{}^a$ $LHSV^{-0.5}$ where $C = C_1-C_9$ hydrocarbon yield weight percent/100; $E =$ activation energy cal/g-mol °C; $A =$ frequency factor; $P_{H_2} =$ absolute hydrogen pressure psia; and LHSV = volume oil feed/volume catalyst–hr.

The frequency factor, activation energy, and hydrogen pressure order (a) are tabulated for each feedstock in Table IV and Figure 1. No attempt was made to study catalyst deactivation rates, but a nominal 48-hr induction period was noted. Where data were generated during an obvious induction period, those data were not used in determining rate constants. However, catalyst loads were used typically for hundreds of hours. The single exception was that used for the Synthoil mid-distillate which could be run for only about 100 hr before a plug in the cooling zone of the reactor would shut down the operation. The plug was ammonium chloride. Opening the reactor to remove the plug necessitated a catalyst change.

The selectivity of the hydrocracking process is dependent upon the severity of operation. The experimental design was such that conversion was largely dependent upon temperature. It was not practical to achieve equivalent conversions by changing liquid hourly space velocity (LHSV). A typical selectivity change with temperature is illustrated in Table V.

In order to compare the four oils, the temperature for equivalent conversion is calculated and the hydrocarbon selectivities corresponding to that temperature are displayed in Table VI.

Table IV. Hydrocracking Mid-Distillates

$$\ln (1 - C) = -Ae^{-E/RT} P_{H_2}{}^a \text{ LHSV}^{-0.5}$$
$$C = C_1\text{–}C_9 \text{ Yield} \qquad P_{H_2} = \text{pressure, psig}$$

	A	E	a
COED			
1st pass	2.23×10^6	38000	1.25
2nd pass	1.50×10^6	38000	1.25[a]
H-Coal			
1st pass	1.04×10^7	38000	1.0[a]
2nd pass	2.20×10^8	43400	1.0[a]
SRC II			
1st pass	2.40×10^2	21000	1.0
2nd pass	4.40×10^4	30500	1.0[a]
Synthoil			
1st pass	1.48×10^7	38000	1.0
2nd pass	1.22×10^7	38000	1.0[a]

[a] Assumed.

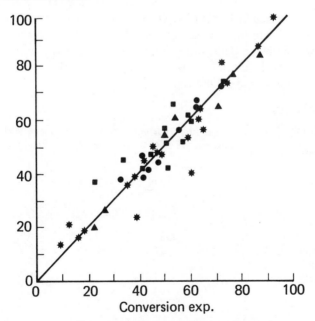

Figure 1. Hydrocracking mid-distillates: (*), COED; (▲), H–Coal; (●), SRC II; and (■), Synthoil.

Table V. Hydrocracking Mid-Distillate COED Oil Yield Patterns (Wt % of Feed)

$2500\ psig;\ LHSV = 0.35$

	$840°F$	$940°F$	$975°F$
Methane	7.1	9.6	14.5
Ethane	8.4	10.8	16.3
Propane	9.3	14.1	17.8
Butane	9.2	11.6	9.8
C_5–C_9			
paraffin	14.8	13.1	6.5
naphthene	30.1	21.1	17.4
aromatic	27.4	26.0	24.7

Gas oil hydrocracking was done on COED and H–Coal gas oils. While these distillates were run from heated feed tanks, pumping difficulties made data acquisition difficult. Selected gas oil hydrocracking data is presented in Table VII.

Hydrotreating. In order to successfully reform a naphtha with the conventional bifunctional reforming catalyst, heteroatoms must be reduced to a level such that they will not deactivate the reforming catalyst. Sulfur and nitrogen levels of well below 5 ppm seem to be required for conventional reforming catalysts while levels below 1 ppm appear necessary for the newer bimetallics. The assumption was made that, if sulfur and nitrogen levels could be reduced to specification, the chemical oxygen levels would follow.

Table VI. Hydrocracking Hydrocarbon Selectivity Mid-Distillates on Harshaw HT-400-E (Wt % of Feed)

2500 psig
0.75 vol oil/hr—vol catalyst
60% wt yield to C_1–C_9

	COED	H–Coal	SRC II	Synthoil
Temperature (°F)	938	961	945	941
Hydrogen	−3.2	−2.9	−4.7	−4.4
Methane	9.5	7.5	5.2	6.2
Ethane	10.3	9.9	8.5	9.4
Propane	13.6	11.8	8.0	9.8
Butane	11.0	9.6	8.3	10.1
C_5–C_9				
paraffin	11.9	12.0	9.0	11.6
naphthene	20.1	21.7	30.7	25.3
aromatic	23.7	27.5	30.4	27.6
H/C atomic ratio product	2.00	1.77	1.72	1.82

**Table VII.   Hydrocracking Hydrocarbon Selectivity
Gas Oil on Harshaw HT-400-E (Wt %)**

2500 psig

	COED	H-Coal
Temperature	932°F	964°F
LHSV vol/vol–hr	1.45 hr^{-1}	1.3 hr^{-1}
Hydrogen	−2.1	−2.9
Methane	6.8	6.0
Ethane	5.7	7.8
Propane	6.2	9.4
Butane	4.0	6.5
C_5–C_9		
paraffin	6.6	7.4
naphthene	8.2	8.9
aromatic	6.7	10.8
C_{10}–650°F	55.8	43.2
Conversion of 650°F+	78%	75%
Feed properties		
carbon	89.0	89.7
hydrogen	10.6	9.22
oxygen	0.25	0.855
nitrogen	0.090	0.083
sulfur	0.009	0.17
specific gravity	0.986	1.01

Hydrotreating experiments were conducted on seven naphthas. The range of conditions under which experiments were conducted is as follows: 500–2000 psig of pressure; an LHSV of 1–4 vol/hr–vol catalyst at 650°–750°F. The catalyst was supported nickel–molybdenum (Ni–Mo), Harshaw HT-100-E, or Cyanamid HDS-9A. Ni–Mo catalysts were used for the presumed nitrogen removal activity.

Several problems were encountered during the hydrotreating processing studies including analytical difficulties and sulfur contamination from the presulfiding procedure. Also, both hydrogen sulfide and water must be removed quantitatively from the product to avoid a room-temperature reaction which results in a high organic sulfur analysis. These effects would tend to give the appearance of a less active catalyst and make the experimental results more difficult to interpret.

The hydrotreating data are summarized in Table VIII. The SRC II straight-run naphtha contained considerably higher levels of heteroatoms than either of the other two straight-run naphthas, and appeared to be the most difficult to hydrotreat. Care must be taken when comparing the COED straight-run naphtha with the other straight-run naphthas since that naphtha is the product of a rather severe fixed-bed hydrogenation in the COED process.

Table VIII. Hydrotreating Naphthas (ppm)

	Sulfur	Nitrogen	Oxygen
COED			
Straight Run			
Feed	40	460	3400
1st Pass	17	84	NA
2nd Pass	< 2	6	280
Hydrocrackate			
Feed	41	< 1	500
Product	< 1	< 1	460
H–Coal			
Straight Run			
Feed	2600	470	3900
1st Pass	60	33	400
2nd Pass	< 4	5	400
Hydrocrackate			
Feed	100	8	300
Product	< 2	2	< 50
SRC II			
Straight Run			
Feed	5400	1130	14100
1st Pass	860	160	2540
2nd Pass	40	< 1	< 70
Hydrocrackate			
Feed	7	160	260
Product	4	< 1[a]	230
Synthoil			
Hydrocrackate			
Feed	30	50	300
Product	< 2	< 1	< 80

[a] Acid wash removed 49 ppm of nitrogen.

Heteroatom levels in the hydrocrackate naphthas could be reduced to acceptable levels in a single stage at relatively severe conditions (1000–1500 psig, 1–2 LHSV). The exception was the SRC II hydrocrackate naphtha where the nitrogen level approached an irreducible 50 ppm. The resistant nitrogen species appear to be alkyl-substituted pyridines; however, they were readily removed with an acid wash. Nitrogen data accumulated during the hydrocracking process from which this naphtha was derived indicate that, at 2500 psig, the hydrocrackate liquid contained 25 ppm of nitrogen; while at 1500 psig the recovered liquid contained 500 ppm.

While it may be possible, with the improved techniques and an acid wash, to hydrotreat the straight-run naphthas in a single stage, the conceptual design resulting from this study will contain a two-stage naphtha hydrotreater in order to assure heteroatom removal to an acceptable level. Hydrocrackate naphthas will feed to the second stage.

Reforming. The hydrotreated naphthas were reformed over a conventional platinum reforming catalyst in an attempt to maximize aromatics. The catalyst was Cyanamid AERO PHF-4 (0.3% Pt, 0.6% Cl). The intent was to operate the reformer at constant conditions in order to better compare naphthas. By operating at severe conditions, the expected hydrocracking activity of the catalyst would tend to purify the aromatics by selectively cracking away the paraffins. If the resultant reformate had a suitably high aromatic content, it could be fed directly to a hydrodealkylator.

An alternate course, which was not pursued, would be to operate the reformer at less severe conditions where naphthene conversions are high but where isomerization and hydrocracking are less. The resultant reformate could be extracted with raffinate going to a naphtha cracker for olefin production while the extract is hydrodealkylated.

The first two naphthas were run at 250 psig. The initial activity was very encouraging, but the activity decline was quite rapid. Since the reactor system was not provided with regeneration capability, the remainder of the naphthas were reformed at 500 psig. At that pressure, clean petroleum naphthas have been run for over 1000 hr with very little activity loss.

The balance of the naphthas were limited in quantity, being the net liquid recovered from a series of experiments, and therefore permitted run lengths of only two to five days. Activity losses were apparent even during these short runs. The rate of activity loss after the first day was virtually the same for all five naphthas run at 500 psig.

The first-day data for the seven naphthas are presented in Tables IX and X. The reformates are notable for the high aromatic content.

Table IX. Reforming Coal-Derived Naphthas: Straight-Run[a]

	COED		H–Coal		SRC II	
	Feed	*Product*	*Feed*	*Product*	*Feed*	*Product*
Temperature		955°F		968°F		960°F
Pressure (psig)		250		250		500
LHSV vol/hr–vol		2.0		1.6		2.3
Hydrogren prod. scf/bbl		1160		1095		1380
Wt (%)						
C_1–C_3		7.5		5.0		7.9
C_4–C_5		5.2		4.1	1.7	10.4
C_6–C_9						
paraffin	10.9	4.0	13.6	6.1	13.9	12.6
naphthene	50.1	0.9	41.3	2.0	71.5	4.4
aromatic	28.7	73.4	33.9	72.0	7.2	60.8
C_{10}+	10.3	6.0	10.2	5.5	7.8	4.0

[a] Catalyst = Cyanamid AERO PHF-4; Hydrogen ratio = 4000 scf/bbl once through.

Table X.   Reforming Coal-Derived

COED

	Feed	Product
Temperature		968°F
Pressure (psig)		500
LHSV vol/hr–vol		1.9
Hydrogen prod. scf/bbl		975
Weight (%)		
C_1–C_3		6.2
C_4–C_5	0.4	5.8
C_6–C_9		
paraffin	13.3	6.4
naphthene	35.7	2.2
aromatic	33.4	70.2
$C_{10}+$	17.2	6.6

[a] Catalyst = Cyanamid PHF-4; hydrogen ratio = 4000 scf/bbl once through.

With the possible exception of the SRC II straight-run naphthas, these C_6–C_9 reformates could be fed to a hydrodealkylator without first being extracted. However, it must be noted that hydrocracking, as evidenced by paraffin conversion, was not nearly as active as expected. In the event that a coal-derived naphtha contained a substantial portion of paraffin, particularly C_6 paraffin, the aromatic content of the resultant reformate would be significantly less.

Summary

The hydroprocessing data for the COED coal-derived synthetic crude oil has been reduced to the point where the chemicals available from the distillate fractions of that oil can be determined (see Table XI).

Table XI.   Chemical Weight Percent Yields from COED Distillates

	Naphtha[a]	Mid-distillate[b]	Gas Oil[c]
Hydrogen	−0.9	−6.3	−7.8
Methane	20.4	20.0	22.9
LPG	20.2	53.3	52.7
Benzene	53.5	31.4	30.8
Liquid fuel	6.8	1.4	1.3
Value/100 lb distillate (1980)	$10.43	$7.29	$7.12

[a] Hydrotreat; reform; hydrodealkylate.
[b] Recycle hydrocrack; hydrotreat; reform; hydrodealkylate.
[c] Recycle hydrocrack gas oil; recycle hydrocrack mid-distillate; hydrotreat; reform; hydrodealkylate.

Naphthas: Hydrocrackate[a]

	H–Coal		SRC II		Synthoil	
	Feed	*Product*	*Feed*	*Product*	*Feed*	*Product*
		968°F		963°F		963°F
		500		500		500
		2.1		2.1		2.1
		1380		900		1150
		6.4		3.5		7.5
		4.8	0.5	2.9		5.6
	10.8	7.0	6.9	4.5	10.1	8.3
	36.9	1.9	25.4	2.1	33.1	2.4
	35.0	74.6	55.1	78.7	36.0	68.4
	17.2	5.2	12.1	8.3	20.7	7.8

The hydrocarbon values are derived from those published by Spitz and Ross (1). Hydrogen was estimated at $2.10/mcf from a methane reformer operating on $3.25/mmBtu gas.

With the fuel value of the feedstock ranging from $6.25 to $6.75 per 100 lb, it would appear that mid-distillate processing could be only marginally attractive, while gas oil processing, with the added processing charges, is likely to be unattractive.

Acknowledgments

The authors thank the U.S. Energy Research and Development Administration without whose financial support this project would not have been accomplished. We also gratefully acknowledge the many people in The Dow Chemical Company who cooperated and assisted in the completion of this work.

Literature Cited

1. Spitz, P. H.; Ross, G. N. "What Is Feedstock Worth?" *Hydrocarbon Process.* **April, 1976.**

RECEIVED June 28, 1978.

Addendum

Subsequent to the appearance of this chapter in the symposium on Refining of Synthetic Crudes, August 28–September 2, 1977, this work has been published in great detail in the following reports to the Energy Research and Development Administration. These reports are available through the National Technical Information Service (NTIS), Springfield, VA 22161.

FE-1534-44   Chemicals from Coal
                  Interim Report for FMC Corporation
                  COED Pyrolysis Process Western Kentucky
                  Syncrude

FE-1534-47   Chemicals from Coal
                  Interim Report for HRI H–Coal

FE-1534-48   Chemicals from Coal
                  Interim Report for USBM Synthoil

FE-1534-49   Chemicals from Coal
                  Interim Report for Solvent-Refined
                  Coal II

FE-1534-50   Chemicals from Coal
                  Final Report based on HRI H–Coal
                  Product

Hydrorefining of Flash Pyrolysis Coal Tar

S. C. CHE, S. A. QADER[1], and E. W. KNELL

Occidental Research Corporation, P.O. Box 19601, Irvine, CA 92713

*The hydrorefining processibility of flash pyrolysis subbitu-
minous coal tar was studied in both batch autoclave and
continuous trickle-bed reactors. Five types of commercial
hydrotreating catalysts were sulfided and precoked to give a
catalyst containing 10% coke deposition. These catalysts
were used to hydrorefine coal tar in a series of screening
tests using the autoclave to determine the most suitable catalyst. Nickel–molybdenum on alumina was the most suitable
catalyst based on the efficiency of heteroatom removal and
hydrogen consumption with cobalt–molybdenum on alumina
being a close second. The trickle-bed experiment using
cobalt–molybdenum alumina catalyst provided confirmation
of the processibility of coal tar in a continuous operation.*

Occidental Research Corporation (ORC) has been developing a pyrolysis process for the production of liquid fuels from coal. In the
ORC flash pyrolysis process finely divided coal is heated at approximately
10,000°F/sec causing the coal particles to decompose very rapidly to
gas, liquid, and char. A very high yield of liquids (35% from Kentucky
bituminous coal), about twice the Fischer assay (16%), can be obtained
because of the short reaction time, minimizing degradation of the tar (1).
A 3-ton/day Process Development Unit (PDU) is presently being operated under DOE Contract EX76-C-01-2244. The process flow schemes and
the PDU flow diagram have been given elsewhere (1, 2).

Liquids from the flash pyrolysis of coal are heavy and viscous. The
hydrogen-to-carbon atomic ratio is about 1.1. They are high in asphaltenes and preasphaltenes, which render them only partially distillable.

[1] Present address: Jet Propulsion Laboratory, California Institute of Technology,
4800 Oak Grove Dr., Pasadena, CA 91103.

0-8412-0456-X/79/33-179-159$05.00/1
© 1979 American Chemical Society

Table I. Properties of

Process Coal	H–Coal (3) bituminous	Synthoil (3) bituminous
ASTM D-1160 distillation		
10%	569°F	473°F
50%	759	715
70%	963	890
% Residue	32	31
Sp gr, °API	—	−2.9
Solubility classification (wt %)		
oils	64.1	56.7
asphaltenes	13.3	29.2
preasphaltenes	16.7	13.2
pyridine insolubles	6.0	1.9
Ultimate analysis (wt %)		
carbon	89.00	87.62
hydrogen	7.94	7.97
nitrogen	0.77	0.97
sulfur	0.42	0.43
oxygen (diff)	1.87	3.02
Atomic H/C	1.07	1.09

They do not differ markedly from coal-derived liquids from a variety of other processes as given in Table I (1, 3, 4).

The liquids require a hydrorefining step to stabilize their reactive properties, to reduce the asphaltenes and preasphaltenes, to reduce sulfur, nitrogen, and oxygen, and to make the liquids more distillable. The extent of hydrorefining depends on the end use of liquids—fuel oil or chemical feedstocks. The objective of this work is to evaluate the hydro-refining processibility of ORC flash pyrolysis coal tar as a part of the tar characterization task. Results of the initial phase of catalyst screening tests are reported in this chapter.

Experimental

Wyoming subbituminous coal, whose analyses are given in Table II, was processed in both the PDU and a bench-scale reactor (BSR). The PDU is a 3 ton/day system while the BSR was designed for 3 lb/hr. The properties of tar produced in these two different systems under similar conditions were almost the same as shown in Table III.

The tar used for this study was obtained from a series of BSR runs. The liquids were dissolved in acetone to filter off carried-over char fines. Acetone was stripped off in a one-plate distillation system. The analyses of this mixture of BSR tar are given in Table IV. The feed was stored at −30°C, under a nitrogen blanket to minimize degradation.

Coal-Derived Liquids

	SRC (3) bituminous	COED (4) bituminous	ORC bituminous	ORC subbituminous
	750°F	615°F	510°F	550°F
	950	777	720	898
	—	1028	960	—
	80	30	32	50
	−11.9	−4.0	−5	−10.7
	14.3	55.0	55.1	42.1
	28.4	28.1	24.9	29.7
	47.3	16.8	19.7	27.7
	9.9	0.1	0.3	0.5
	87.93	79.6	81.5	79.75
	5.72	7.1	7.1	7.50
	1.71	1.1	1.4	1.41
	0.57	2.8	2.1	0.64
	4.07	8.5	7.9	10.70
	0.76	1.07	1.05	1.13

Five types of commercially available petroleum hydrotreating catalysts were evaluated. The properties of catalysts are listed in Table V. All catalysts were presulfided before being coked.

Coking Experiments. The sulfided catalysts were first deposited with coke before catalyst screening. For these experiments, 120 g of PDU tar and 40 g of sulfided catalyst were charged into a 1-L Magnedrive

Table II.   Analyses of Subbituminous Coal[a]

Moisture (wt %) (as fed)	11.6

Ultimate analysis (wt %) (dry basis)

carbon	69.1
hydrogen	5.0
nitrogen	1.2
sulfur	0.7
oxygen (by diff)	18.3
ash	5.7

Proximate analysis (wt %) (dry basis)

volatile matter	41.8
fixed carbon	52.5
ash	5.7
heating value, cal/g	6,360

[a] Monarch Seam, Big Horn, Wyoming.

Table III. Analyses of Tar

	PDU	BSR
Pyrolysis condition		
temperature (°F)	1050	1000
residence time (sec)	2.5	3.0
Tar properties		
gravity (°API)	−8.6	−8.1
viscosity at 250°F (cP)	115	135
Tar analysis		
solubility class (wt %)		
oil[a]	54.9	56.4
asphaltenes[b]	28.0	27.4
preasphaltenes[c]	14.8	16.1
char[d]	2.2	0.1
Ultimate analysis (wt %)		
carbon	79.8	79.1
hydrogen	7.3	7.8
nitrogen	1.1	1.0
sulfur	0.6	0.4
oxygen (diff)	10.8	11.6
Distillation[e]		
IBP (°F)	430	500
50% distilled (°F)	730	800

[a] Oil: pentane soluble.
[b] Asphaltenes: toluene soluble, pentane insoluble.
[c] Preasphaltenes: pyridine soluble, toluene insoluble.
[d] Char: pyridine insoluble.
[e] ASTM D-1160.

Table IV.   Analyses of Hydrorefining Feed

Ultimate analysis (wt %)

ash	0.12
carbon	79.48
hydrogen	6.84
nitrogen	1.36
sulfur	0.45
oxygen (by diff)	11.75
Sp gr, @ 60°F/60°F	1.177
°API	−11.3

Solubility classification (wt %)

oil[a]	46.8
asphaltenes[b]	33.7
preasphaltenes[c]	19.3
pyridine insolubles	0.2

Distillation[d]

IBP	135°F
5%	443
10%	488
20%	566
30%	680
40%	747
50%	825
62%	850

Residue: 39.6 wt %

[a] Oil: pentane soluble.
[b] Asphaltenes: toluene soluble, pentane insoluble.
[c] Preasphaltenes: pyridine soluble, toluene insoluble.
[d] ASTM D-1160.

Table V. Catalyst Properties

Catalyst	Shell 244	Shell 214	Shell 344	Shell 324	Shell 454
Composition (% wt)					
cobalt	2.2	—	2.4	—	—
nickle	—	2.2	—	2.7	5.2
molybdenum	9.0	8.9	10.0	13.7	—
tungsten	—	—	—	—	24.0
Support	Alumina	Alumina	Silicated Alumina	Silicated Alumina	Alumina
Mercury pore-size distribution, % V of PV					
< 50 Å	0	0	1.9	1.0	1.3
50–70 Å	1.8	1.0	11.8	10.0	12.8
70–100 Å	11.9	10.6	36.8	26.0	61.7
100–150 Å	60.2	66.2	38.1	61.7	22.9
150–350 Å	25.0	20.7	9.3	1.0	0.9
> 350 Å	1.1	1.5	2.2	0.5	0.4
Av pore size, Å	133	132	98	109	93
Physical properties					
compacted bulk density (kg/M^3)	720	690	770	850	1060
attrition index	99.5	99.5	99.0	99.0	99.0
dry H_2O PV (cc/g)	0.63	0.63	0.59	0.49	0.38
surface area (M^2/g)	160	159	192	162	157
Size (extrudates) (mm)	1.6	1.6	1.6	1.6	1.6

autoclave. The autoclave was pressurized to 2000 psi of H_2 before heating up to 400°C. After the desired reaction time, the reaction was quenched. Catalysts were separated from tar, washed with toluene, and dried. A portion of catalyst, approximately 10 g, was withdrawn for coke determination by burning the catalyst at 480°C. After a bulk quantity of catalyst was prepared, the coked catalysts were used for catalyst comparison experiments.

Catalyst Screening. The autoclave was modified to allow preheating and tar injection under pressure for the catalyst comparison runs. A high-pressure bomb made of stainless steel was used as an injection vessel. The bomb was equipped with a temperature-controlled heater to keep the tar at 150°C. A pressure head higher than the autoclave-system pressure is maintained on the tar so that the tar can be injected into the preheated autoclave after the isolation valve between the bomb and autoclave was open.

The autoclave was charged with 45 g of precoked catalyst and pressurized to 1800 psi with hydrogen. BSR tar, approximately 120 g, was injected into the autoclave when the internal temperature was stabilized at 375°C. The reaction system was quenched after 1 hr of reaction. The product gas was vented and analyzed. The liquid products were separated from the catalyst and analyzed for carbon, hydrogen, nitrogen, sulfur, asphaltenes, and preasphaltenes. Coke deposition on the catalysts was analyzed again. Catalyst performance was evaluated based on liquid-product properties.

The specific surface area was determined by a Micromeritics Model 2200 high-speed, surface-area analyzer using nitrogen as the adsorbate. The pore volume was determined by the mercury penetration method on a Micromeritics Model 900/910 series porosimeter.

Results and Discussion

The study of catalytic reactions using a batch autoclave reactor has been criticized for long heat-up times and nonsteady-state catalyst activity. This catalyst-comparison study was carried out by injecting coal tar into a preheated autoclave containing precoked catalyst and hydrogen in order to eliminate long heat-up times. Nevertheless, the inherent differences in results between an autoclave and a trickle-bed reactor, which is widely used for hydrorefining of heavy liquids, exist still (5).

Time Dependence of Coke Deposition. The results of coke deposition as a function of time are given in Table VI. Approximately 1 hr was required for the reactants to reach 400°C. The reaction time is the duration at 400°C. At zero reaction time, the coke level was approximately 10%. It varied slightly with time. The 1-hr reaction time was chosen based on the system pressure which levels off at 50 min after the system reaches 400°C as shown in Figure 1. A reaction time of 2 hr gave an undesirably high coke level of 21.8%. A 1-hr reaction time gave approximately 10% coke deposition on the catalyst. A 10% coke deposition on the catalyst is high for a petroleum hydrorefining process. However, such

Table VI.   Coke on Catalyst[a]

Coke on Catalysts (wt %)

Reaction Time[b] (min)	Co–Mo on Al_2O_3	Ni–Mo on Al_2O_3	Co–Mo on SiO_2– Al_2O_3	Ni–Mo on SiO_2– Al_2O_3	Ni–W on Al_2O_3
0	11.7	7.9	9.9	11.7	10.5
15	—	—	12.0	—	8.6
30	10.8	9.3	11.9	9.4	10.4
45	12.8	10.2	—	9.1	9.7
60	10.0	9.7	10.9	11.9	—

[a] 400°C, 2000 psi H_2, 40 g sulfided catalyst, 120 g tar.
[b] Reaction time: time at 400°C. Approximately 1 hr was required for the reactants to reach 400°C.

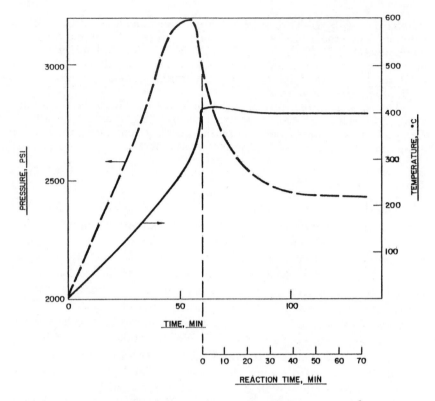

Figure 1.   Pressure and temperature vs. reaction time. Initial pressure: 2000 psi; 40 g sulfided catalyst and 120 g tar.

high coke levels have been observed in the coal-conversion processes. For example, the H–coal process reported that its operating catalyst normally has 10–20% carbon content (6).

Temperature and Pressure Dependence of Coke Deposition. The initial hydrogen pressure and reaction temperature were varied to determine their effects on coke deposition. The results are illustrated in Figure 2. The system temperature has more effect on the coke deposition than the hydrogen pressure. An increase of reaction temperature severely increases coke deposition on the catalyst. Decreasing hydrogen pressure to a lesser extent increases coke deposition. The results also indicate that 2000 psi of H_2 and 400°C were proper for coking catalysts.

Physical Properties of Coked Catalysts. Surface areas for a series of Shell 244 (cobalt–molybdenum (Co–Mo) on alumina) catalysts varying from 0 to 22% coke were determined. The surface area is inversely proportional to coke deposition as shown in Figure 3. The catalysts with 10% coke deposit lose approximately 20% of their original surface areas.

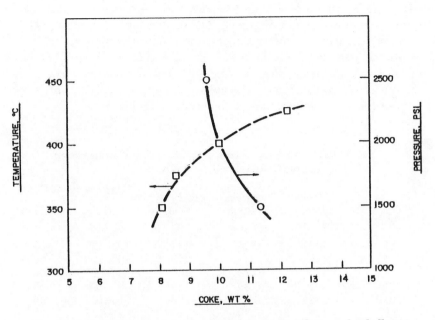

Figure 2.   Coke vs. temperature and pressure. Catalyst used: Shell 244 (Co–Mo on alumina).

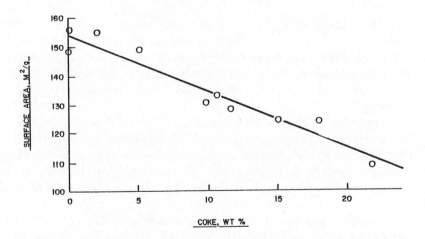

Figure 3.   Effect of coke on surface area of catalyst. Catalyst used: Shell 244 (Co–Mo on alumina).

Pore volumes and average pore sizes for these catalysts are given in Table VII. The relationship of pore volumes and pore sizes for the coked catalysts are illustrated in Figure 4. The smaller pores (less than 200 Å) were lost significantly during the catalyst-deactivation process. The loss of surface area owing to coke was from the loss of small pores. Despite the deactivation owing to coke, the 10% coked catalysts used in the catalyst screening runs have a moderate surface area of 130 m²/g and 77% of the available pores are in the range of 75–175 Å.

Table VII.   Surface Areas and Pore-Size Distribution of Coked Shell 244 Catalyst

Coke (wt %)	Surface Area (M²/g)	Pore Volume (cc/g)	Av Pore Size (Å)	Most Frequent Pores Size (Å)	% Distribution
0	148	0.55	147	100–125	38
2.2	155	0.51	115	100–125	62
5.1	149	0.48	100	100–125	56
9.9	131	0.39	119	75–100	39
10.9	134	0.42	126	75–100	40
11.7	128	0.39	122	75–100	39
15.1	124	0.30	105	100–125	31
18.0	125	0.27	105	75–100	29
21.8	109	0.26	110	100–125	32

Deactivation owing to metal poisoning was not determined. The Wyoming coal contained significant amounts of iron, calcium, magnesium, sodium, and titanium. Some of these metals would exist in the coal tar and be deposited on the catalyst. However, concentrations were too low to be significant compared with the aged catalyst from a trickle-bed operation.

The coked catalysts lost less than 2% coke during the heat-up process in the catalysts screening experiments. The increase of surface areas and pore volumes was not significant. The decrease of coke could be offset when the coal tar is injected into the autoclave. At the end of the experiment, the coke deposit on the catalyst was the same as the original level.

Catalyst Screening Results. A comparison of the catalysts' performance is given in Table VIII. Shell 214, (nickel–molybdenum (Ni–Mo)) on alumina, is the best among the five types of catalysts tested with Shell 244, Co–Mo on alumina being nearly as good. It achieved high removal of heteroatoms with the least hydrogen consumption. The

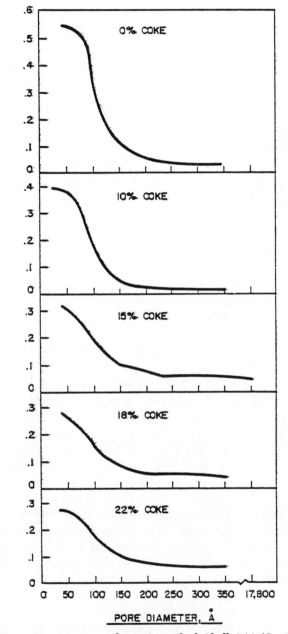

Figure 4.   Pore size vs. pore volume for coked Shell 244 (Co–Mo on alumina)

Table VIII.   Comparison

Catalyst	Pyrex Glass Beads	Shell 244
Composition	—	Co–Mo on Al$_2$O$_3$
H/C of liquid product	0.877	1.394
ΔAPI		28.1
% Removal		
sulfur		86.7
nitrogen		59.6
asphaltenes[b]		95.5
preasphaltenes[c]		99.5
Wt % of carbon-formed gases	2.7	12.8
H$_2$ consumption (scf/bbl)	124	3200
distribtuion (%)		
hydrogenation		58.5
heteroatom removal		30.2
H/C gas formation		11.3

[a] Run condition: 1800 psi H$_2$ (cold) and 380°C, 400 rpm, 1 hr. Tar was injected in autoclave when system reaches 300°C. Feed: H/C = 1.033; −11 °API.

acidity provided by silica–alumina, Shell 324 and 344, lowered the hetero-atom removal while increasing hydrocarbon gas formation. This could be because of nitrogen poisoning of acidic sites. Nickel–tungsten (Ni–W) removed sulfur more effectively than nitrogen. The hydrogen consumption distribution and gas formation indicated that Ni–W has high hydrocracking activity.

The pyrex glass beads only provided thermal cracking activity. The liquid products were more viscous, heavier, and hydrogen-deficient than the original coal tar. It also provided evidence that the upgrading of liquids is mainly catalytic.

Removal of asphaltenes and preasphaltenes was easier than hetero-atoms. At an increase of a hydrogen-to-carbon atomic ratio by 0.16, they were 80% removed. A comparison can be made with Synthoil process product of hydrogen-to-carbon atomic ratio equal to 1.04 which was hydrotreated to hydrogen-to-carbon atomic ratio equal to 1.24. The asphaltenes and preasphaltenes were 77% and 99% removed, respectively (7). Squires (8) concluded that the preasphaltenes can be converted to asphaltenes and oils with very little consumption of hydrogen. Asphaltenes are the major consumers of hydrogen. Although Squires' conclusions were based on donor-solvent coal liquefaction, similar results were reported

of Catalysts' Performance[a]

Shell 214	Shell 344	Shell 324	Shell 454
Ni–Mo on Al_2O_3	Co–Mo on SiO_2–Al_2O_3	Ni–Mo on SiO_2–Al_2O_3	Ni–W on Al_2O_3
1.270	1.346	1.197	1.057
27.0	22.0	22.1	15.6
84.4	66.7	68.9	80
51.5	30.9	24.3	nil
93.5	83.4	79.2	72.2
97.9	96.4	80.8	79.9
0.8	5.4	7.7	7.1
2460	3000	2630	1360
49.9	54.1	46.5	9.1
41.0	34.5	36.6	59.9
9.1	11.4	16.9	31.0

[b] Asphaltenes: Toluene soluble, pentane insoluble.
[c] Preasphaltenes: Pyridine soluble, toluene insoluble.

by deRosset in the trickle-bed hydrotreating of various coal-derived liquids (7,9) and observed in our experimental results. In view of these facts, the emphasis of catalyst selection should be based on the degree of hydrogen consumption and heteroatom removal.

Trickle-Bed Hydrorefining. A 1-in. ID trickle-bed reactor containing 80 g of calcined Shell 244 catalyst, Co–Mo on alumina, was used to hydrotreat the flash pyrolysis subbituminous coal tar from the PDU. The pyrolysis liquid was filtered prior to trickle-bed hydrotreating. It contained 0.2% pyridine-insoluble material. The test was carried out at 2500 psi, 400°C, and 1.0 weight hourly space velocity (WHSV). The results are illustrated in Figure 5. The liquid products were lighter compared with those of the autoclave tests. They contained less heteroatoms, asphaltenes, preasphaltenes, and more hydrogen because of more severe processing conditions.

The formation of gas was low. From the carbon balance, only 3% of the total carbon formed gaseous products. The hydrogen consumption was approximately 3100 scf/bbl, of which nearly 80% was consumed in liquid hydrogenation. This preliminary result demonstrated that the flash pyrolysis coal tar derived from a subbituminous coal is amenable to hydrorefining.

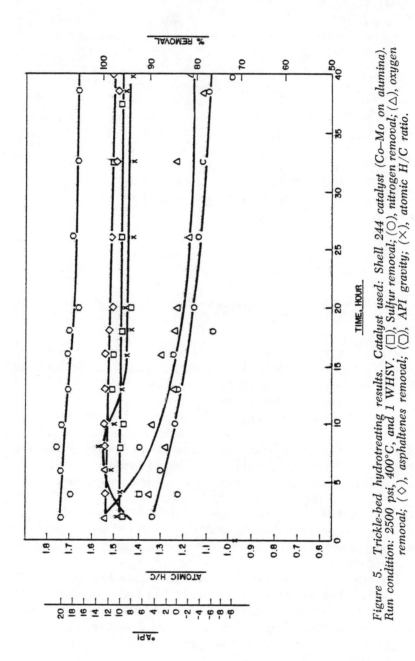

Figure 5. Trickle-bed hydrotreating results. Catalyst used: Shell 244 catalyst (Co—Mo on alumina). Run condition: 2500 psi, 400°C, and 1 WHSV. (□), Sulfur removal; (○), nitrogen removal; (△), oxygen removal; (◇), asphaltenes removal; (○), API gravity; (×), atomic H/C ratio.

Conclusion

The results of this study showed that flash pyrolysis subbituminous coal tar can be hydrorefined. Ni–Mo supported on alumina was the most effective catalyst for hydrorefining flash pyrolysis subbituminous coal tar among the five types of catalysts tested. The removal of asphaltenes and preasphaltenes was easy to achieve. Catalyst selection should emphasize heteroatoms' removal efficiency and hydrogen consumption. The catalyst life is an equally important selection criterion although this program has not completed catalyst-life studies. It will be determined in the future.

Acknowledgment

This work was supported by DOE under contract EX76-C-01-2244.

Literature Cited

1. Che, S. C.; Dolbear, G. E.; Longanbach, J. R. *Preprints of the Coal Chemistry Workshop,* Stanford Research Institute, Menlo Park, CA, Aug. 1976, p. 40.
2. Knell, E. W.; Gethard, P. E.; Shaw, B. W.; Green, N. W.; Qader, S. A. "Flash Pyrolysis Coal Liquefaction Process Development," FE-2244-4, Quarterly Report for July–Sept. 1976, Occidental Research Corp., Irvine, CA, 1976, pp. 35–41.
3. Bendoraitis, J. G.; Cabal, A. V.; Callen, R. B.; Stein, T. R.; Voltz, S. E. "Upgrading of Coal Liquids for Use as Power Generation Fuels," Phase I Report, Research Project 361-1, Electric Power Research Institute, Palo Alto, CA, 1976, pp. 28–34.
4. Scotti, L. J.; McMunn, B. D.; Greene, M. I.; Merrill, R. C.; Schoemann, F. H.; Terzian, H. D.; Hamshar, J. A.; Domina, D. J.; Jones, J. F. "Char Oil Energy Development," R&D Report No. 73—Interim Report No. 2, Contract No. 14-32-0001-1212, Office of Coal Research, Washington, D.C., 1974, p. 57.
5. Satterfield, C. N. *AIChE J.* **1975,** *21* (2), 209.
6. Kang, C. C.; Johnson, E. S. *Am. Chem. Soc., Div. Fuel Chem., Prepr.* **1976,** *21* (5), 32.
7. deRosset, A. J. "Characterization of Coal Liquids—Upgrading Synthoil," FE-2010-03, UOP Inc., Des Plaines, IL, 1976.
8. Squires, A. M. "Reaction Paths in Donor Solvent Coal Liquefaction," presented at *Conference on Coal Gasification/Liquefaction, Moscow, USSR, October, 1976.*
9. deRosset, A. J.; Tan, G.; Gatsis, J. G.; Shoffner, J. P.; Swenson, R. F. "Characterization of Coal Liquids," FE-2010-09, UOP Inc., Des Plaines, IL, 1977.

RECEIVED June 28, 1978.

An Investigation of the Activity of Cobalt-Molybdenum-Alumina Catalysts for Hydrodenitrogenation of Coal-Derived Oils

M. M. AHMED[1] and B. L. CRYNES

School of Chemical Engineering, Oklahoma State University, Stillwater, OK 74074

FMC oil and raw anthracene oil (coal-derived liquids) have been hydrotreated in trickle-bed reactors using two commercially available cobalt–molybdenum–alumina catalysts. Results indicated that denitrogenation tends to be more difficult with an increase in the boiling range of the feed oil. Distillation of feed and product oils revealed significant differences in the distribution of nitrogen in the various fractions. Under relatively mild processing conditions (371°C and 1500 psig), a bidispersed catalyst was observed to better denitrogenate FMC oil than a monodispersed catalyst. However, for the relatively lighter liquid, raw anthracene oil, there was little difference in the activity of the two catalysts. The hydrodenitrogenation activity for both, mono- as well as bidispersed, decreased significantly within 100–150 hr of oil–catalyst contact time. Carbonaceous depositions seem to be the primary cause for catalyst activity decay.

Coal is one of the most abundant fossil fuels in the U.S., and the increasing cost and depleting resources of energy have focused the attention of many researchers on the development of coal conversion technology. Coal liquids from some of the liquefaction processes under consideration may require secondary hydrotreatment for heteroatom removal to upgrade these liquids.

[1] Current address: Westinghouse Research & Development Center, Pittsburgh, PA 15235.

0-8412-0456-X/79/33-179-175$05.00/1

Cobalt–molybdenum–alumina (Co–Mo–alumina), nickel–molybdenum–alumina (Ni–Mo–alumina), and nickel–tungsten–alumina (Ni–W–alumina) catalysts are used commercially for hydrotreating petroleum feedstocks. Relatively little information is available on the hydrodenitrogenation as compared with hydrodesulfurization of both petroleum feedstocks and coal-derived liquids (1). Mechanistic aspects of the hydrodenitrogenation reaction network and the relative performance of commercially available hydrotreating catalysts are being investigated in batch autoclave or continuous-flow microreactor units using model compounds like quinoline, carbazole, acridine, and benzoquinolines (2, 3, 4). Shih et al. (4), using Ni–Mo–alumina, Ni–W–alumina, and Co–Mo–alumina catalysts, observed the hydrodenitrogenation of quinoline to be relatively easier than that of acridine. They reported that Ni-based catalysts were better than Co-based catalysts. Ring saturation followed by the cleavage of the C–N bond were the distinct reaction steps. The hydrogenation activity was affected by the promoter metal (Ni/Co), whereas the cracking activity was more influenced by the source of the catalyst support material (alumina). Shabtai et al. (3) observed steric effects to play an important role in the hydrodenitrogenation of isomeric benzoquinolines. For 7, 8-benzoquinoline, the saturation of the pyridine ring was followed by the cleavage of the C–N bond. However, the pyridine ring saturation was much faster for 5, 6-benzoquinoline and was followed by the hydrogenation of the end benzene ring prior to C–N bond rupture. Flinn et al. (5) have reported the hydrodenitrogenation of model compounds in representative petroleum fractions under industrial processing conditions. They observed radically different denitrogenation rates for compounds boiling in the same range and also reported that the denitrogenation tends to be more difficult with an increase in the boiling point of the petroleum feed.

Commercially, hydrotreatment of petroleum feedstocks is practiced in trickle-bed reactors. Mears (6), Paraskos et al. (7), and Montagna and Shah (8) have reported that the hydrodenitrogenation data for petroleum feedstocks using trickle-bed reactors is represented better by pseudo power law models accounting for the fluid dynamic aspects (liquid holdup/catalyst wetting effects). Satchell (9) has reported that mass transfer effects were negligible in the hydrodenitrogenation of raw anthracene oil (feedstock used in this study) under industrial processing conditions. Sivasubramanian (10), using a trickle-bed reactor setup similar to that of Satchell's, but with a different set of catalysts, observed that the hydrodenitrogenation of raw anthracene oil was better represented by a first-order model accounting for the partial wetting of the catalysts.

In the case of coal-derived liquids a variety of polynuclear aromatic compounds with a broad spectrum of molecular sizes are encountered. For efficient utilization of the catalyst, these molecules should diffuse into the porous structure for availing the active surface. Significant pore diffusional limitations have been observed in the hydrotreatment of petroleum crudes (*1*), but very little is known concerning such effects for coal liquids. The objective of this study has been to investigate the performance of commercially available Co–Mo–alumina catalysts with significantly different support properties in hydrotreating relatively light and heavier coal liquids. Considering the complexity of coal liquids and the importance of trickle-bed reactor fluid dynamics, the activity test data reported in this chapter can speculate more realistically the performance of Co–Mo–alumina catalysts in commercial-scale hydrotreatment of coal liquids than the results from batch autoclave or microreactor studies using model compounds.

Coal Liquid Feedstocks

Experience with the development of the solvent-refined coal (SRC) process had suggested that raw anthracene oil was a suitable coal liquid for initial catalyst studies; hence some data were obtained using this type of oil. A filtered oil obtained from the COED process developed by FMC Corporation also was used in these studies. The properties of these two oils are given in Table I. The total nitrogen concentration in the raw anthracene feed oil was 0.97 wt % and the FMC oil contained 1.13 wt % nitrogen. Both liquids contained negligible quantities of ash materials.

Table I. Feedstocks and Their Properties

	Raw Anthracene Oil (wt %)	FMC Filtered Oil (wt %)
Carbon	90.65	83.0
Hydrogen	5.76	8.35
Sulfur	0.48	0.35
Nitrogen	0.97	1.13
Oxygen[a]	2.13	7.15
Ash	Nil	Nil
API Gravity (60°F)	−7	−4.5
IBP	815°F (193°C)	
90%	815°F (435°C)	

[a] By difference.

Catalyst Properties

Two Co–Mo–alumina catalysts obtained from a commercial vendor as either marketed or special research samples were used in this study. The surface area and pore-size distributions (using the mercury penetration technique) were determined by an independent commercial laboratory. The catalyst properties are given in Table II. Note that the monodispersed (MD) and bidispersed (BD) catalysts have the same metallic composition and are chemically similar.

Table II.   Catalysts and Their Properties

	MD (Extrudate)	BD (Spheroids)
[a]CoO, wt %	3.5	3.5
[a]MoO$_3$, wt %	12.5	12.5
Support	Alumina	Alumina
Pore vol, mL/g	0.463	0.981
Surface area, m^2/g	240 (270[a])	286 (270[a])
Most frequent pore radius, Å	33.0	30 micro, 890 macro
Particle size	8/10 mesh	8/10 mesh

[a] Vendor data.

Experimental System and Procedure

The experimental system used in this study is shown in Figure 1. The trickle-bed reactor was made up of 1.27-cm (½-in) OD, Type 316 stainless steel tubing with 0.32-cm (⅛-in.) OD, centrally located thermowell tubing. The reactor was packed with 8/10 mesh catalyst particles. The nominal reactor temperature was measured by a thermocouple which could traverse axially along the thermowell from the top to the bottom of the catalyst bed. The nominal reactor pressure was measured by a Heise gauge which was located upstream from the reactor.

Hydrogen and feed oil were allowed to combine at the top of the reactor before entering the catalyst bed. Hydrogen was used directly from the vendor supply cylinders on a once-through basis. The feed oil was pumped at a constant preset rate by a Ruska pump. The liquid mass flux used in this study was between 0.028 and 0.2 kg/m^2/sec and these flow rates are well within the range of other trickle-bed reactor studies (9, 10). The gases and liquids leaving the reactor were separated in the sample bombs and the liquid samples were collected by isolating the lower bomb from the rest of the system. The flow rate of the exit gases was monitored using a bubble flow meter.

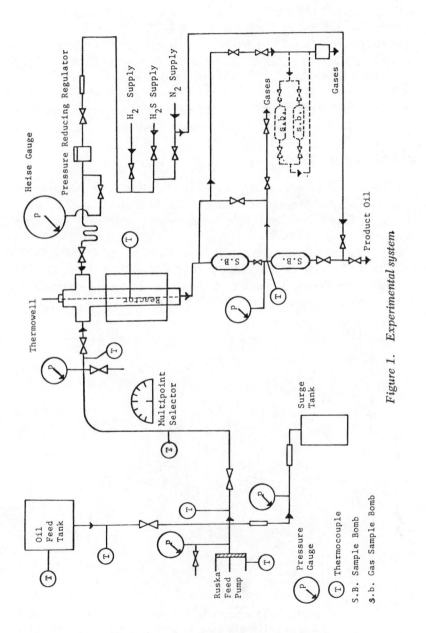

Figure 1. Experimental system

The catalyst pretreatment consisted of calcination and presulfiding steps. The catalyst bed was heated gradually and the temperature was stabilized at 232°C (450°F) for 12 hr with nitrogen flowing through the system. The catalyst then was presulfided at 232°C (450°F) with a 5-vol % hydrogen sulfide-in-hydrogen mixture for 90 min. The system then was flushed with hydrogen for 10 min at a low flow rate. Next, the reactor was heated to within 5.5°C (10°F) of the desired value before bringing hydrogen and feed oil on stream. The temperature and hydrogen flow rate were stabilized at the desired values and the experiments were conducted for 100 to 200 hr of continuous operation.

Results and Discussion

The results from a series of experimental runs using a commercially available BD Co–Mo–alumina catalyst for hydrotreating FMC oil are shown in Figure 2. The hydrodenitrogenation activity of the catalyst decreased gradually at 371°C (700°F) and 427°C (800°F). A rapid decay in the hydrodenitrogenation activity was observed at 454°C (850°F) with the activity level approaching that at 371°C (700°F) within 140 hr. As shown in Figure 3, significant decay in the hydrodenitrogenation activity of the MD catalyst was observed also.

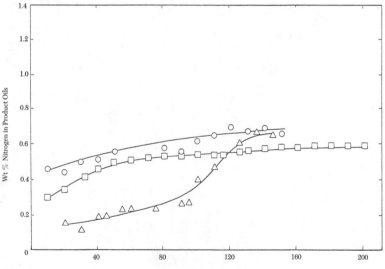

Figure 2.   Catalyst hydrodenitrogenation activity response. Feed: FMC oil; catalyst: BD; pressure: 1500 psig; H_2/oil ratio: 7500 scf/bbl; space time: 2.99 hr (LVHST). (○), 371°C (700°F); (□), 427°C (800°F); (△), 454°C (850°F).

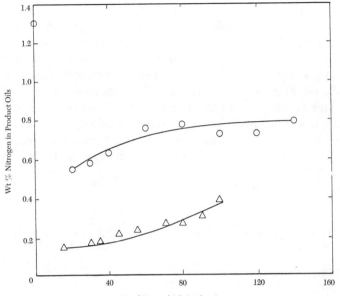

Figure 3. Temperature effect on FMC oil hydrodenitrogenation. Catalyst: MD; pressure: 1500 psig; space time: 2.99 hr (LVHST). (○), 371°C (700°F); (△), 427°C (800°F).

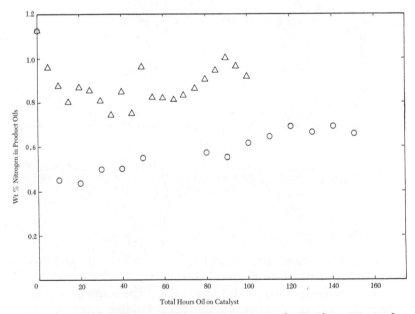

Figure 4. Results for presulfiding–nitrogen removal. Catalyst: BD; feed: FMC oil; pressure: 1500 psig; temperature: 371°C (700°F); space time: 2.99 hr (LVHST). (○), SC–E and (△), SC–A.

Results of two experimental runs conducted under identical conditions except for the presulfiding stage are shown in Figure 4. In Run SC–A less than the stoichiometric requirement of hydrogen sulfide (11) was supplied during the presulfiding operation, whereas the catalyst was presulfided adequately in Run SC–E. The catalyst in Run SC–A achieved some degree of on-stream sulfiding as indicated by an increase in the activity during the first 40 hr of operation. However, the activity level never approached that of the properly treated catalyst in Run SC–E.

Results for raw anthracene oil hydrodenitrogenation are shown in Figure 5 and there seems to be negligible difference in the activity of the two catalysts. The activity decay also seems to be rather small when compared with that for FMC oil.

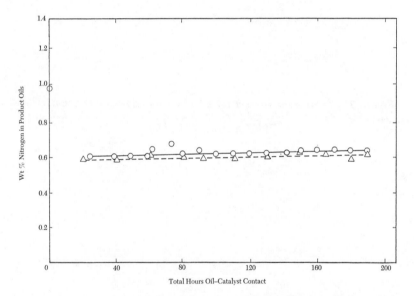

Figure 5.   Hydrodenitrogenation of raw anthracene oil.   Catalyst: (○),
BD, LVHST = 0.925 hr; (△), MD, LVHST = 0.64 hr.   Pressure: 1000
psig; temperature: 371°C (700°F).

General Mechanism

Group-type mass spectrometric (MS) analysis of feedstock and upgraded raw anthracene oil revealed the presence of about 15 aromatic nitrogen-containing compound types (12). The analytical data also showed that the composition of nitrogen-containing aromatics is more complex in the products than in the feed oil. For example, the carbon number distribution for the basic nitrogen-containing compound types are consistent with the presence of 62 and 77 homologs in the feed and product oils, respectively. The molecular weight of nitrogen-containing compounds in these samples varied from 79 to 261. Some representative molecular structures consistent with the molecular formulas of a number of the nitrogen-containing aromatics are given in Table III. These two- and three-ring compounds have boiling points ranging from 232° to 371°C (450°–700°F).

The distillation curves (ASTM D-1160) for the two feed oils and a selected product sample for each are presented in Figure 6. Only 40 vol % of FMC oil boiled to 271°C (520°F) at 50 mm Hg, and approximately 78 vol % of raw anthracene oil boiled in this temperature range. Raw anthracene oil, when hydrotreated at 371°C (700°F), 1000 psig, 1.5 LVHST, and 1500 scf H_2/bbl oil, gave a product, 86 vol % of which boiled to 271°C (520°F). FMC oil hydrotreatment at 371°C (700°F), 1500 psig, 2.99 LVHST, and 7500 scf H_2/bbl yielded a product of which only 54 vol % boiled below 271°C (520°F). Such shifts in the boiling ranges are indicative of the hydrogenation–hydrogenolysis reactions that occur during the hydrotreatment operation.

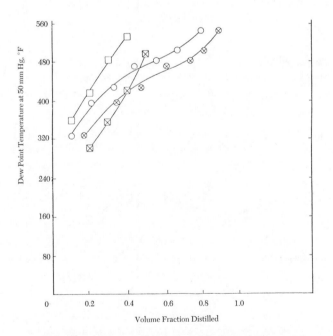

Figure 6.   FMC oil and raw anthracene oil boiling ranges at 50 mm Hg using an MD catalyst. FMC oil: (□), feed; (⊠), product—371°C (700°F), 1500 psig, 2.99 hr (LVHST). Raw anthracene oil: (○), feed; (⊗), product—371°C (700°F), 1000 psig, 1.50 hr (LVHST).

Table III. Some Representative Heterocyclic Nitrogen Compounds in Raw Anthracene Oil

Proposed Name	Formula	Structure	T (°F) at 760 mm Hg (bp)
Indoline	C_8H_9N		442.4–446
Quinoline	C_9H_7N		458.6
Isoquinoline	C_9H_7N		469.4
1,2,3,4-Tetrahydroquino- line	$C_9H_{11}N$		481.1
1,2,3,4-Tetrahydroiso- quinoline	$C_9H_{11}N$		453.2
Carbazole	$C_{12}H_9N$		671
Propylaniline	$C_9H_{13}N$		431.6
O-Ethylaniline	C_8H_4N		408–410
Indole	C_8H_7N		489.2

Table IV. Distribution of Oil as a Function of Boiling Range

| | FMC Oil | | Raw Anthracene Oil | |
	Feed	Product	Feed	Product
Light oil[a] (vol %) ($T < 270°F$)	—	18.5	—	15.6
Middle distillate[a] (vol %) ($270°F < T < 462°F$)	24.8	29.5	43.0	45.0
Heavy ends[a] (vol %) ($462°F < T$)	75.2	52.0	57.0	39.4

[a] Boiling point at 50 mm Hg.

ASTM charts (D2892) are used in the petroleum industry for converting the boiling ranges of petroleum hydrocarbons from any subatmospheric pressure to that at 760 mm Hg. Assuming their applicability, the coal liquid fractions boil (at 50 mm Hg) to 132.2°C (270°F), 238.9°C (462°F) and those at higher temperatures are considered to be light oils, middle distillates, and heavy ends, respectively. The volume distribution of the various fractions in the feed and product oils of this study are summarized in Table IV. The amount of heavy ends in the product oils seem to decrease while there is a slight increase in the middle distillate fractions. The nitrogen distribution in these various fractions is given in Table V. The greatest quantity of nitrogen is in the middle distillate fraction of the raw anthracene oil; however, the product oil contains appreciably less nitrogen in that fraction. There is little change in the nitrogen content of the middle distillate fraction of FMC oil, but the heavy ends have undergone significant denitrogenation.

Table V. Distribution of Nitrogen as a Function of Boiling Range

| | FMC Oil | | Raw Anthracene Oil | |
	Feed	Product	Feed	Product
Light oil (wt % nitrogen) [a]	—	0.093	—	0.044
Middle distillate (wt % nitrogen) [a]	0.189	0.172	0.56	0.139
Heavy ends (wt % nitrogen) [a]	0.941	0.398	0.41	0.207
Total wt % nitrogen [b]	1.13	0.661	0.97	0.39

[a] Wt % nitrogen $= \dfrac{\left(\begin{array}{c}\text{wt \% nitrogen in any fraction}\\ \text{(light, middle, or heavy)}\end{array}\right)\left(\begin{array}{c}\text{wt \% of that}\\ \text{fraction}\end{array}\right)}{100}$.

[b] Wt % nitrogen of total sample before distilling.

The general hydrodenitrogenation reaction mechanism often presented for compounds like quinoline which boils ca. 260°C (500°F) at 760 mm Hg is as follows (2, 4):

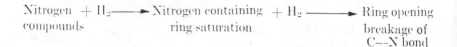

For example: Denitrogenation

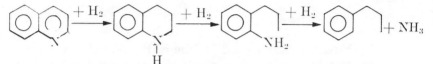

An interesting observation that can be made from the physical properties of the products of the ring-saturation and ring-opening reactions is that these products boil in a range close to that of the parent compound (Table III). Selective hydrogenation–hydrogenolysis of the nonheterocyclic polynuclear aromatics, rather than their heterocyclic counterparts, would result in the concentration of nitrogen in the heavy ends. Contrary to this the results (Figure 6 and Tables IV and V) indicate that the nitrogen content of the heavy ends is reduced considerably, and the product oils contain a significantly lower amount of heavy ends than their respective feed oils. Generally, if the feed is assumed to be composed of a heavy end and a middle distillate fraction represented by **A** and **B**, respectively, then the following general reaction scheme is possible:

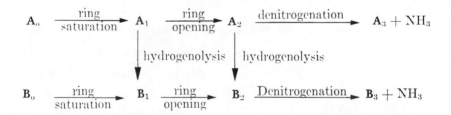

A_0 and B_0 are the initial amounts present in the feed, and A_0, A_1, and A_2 refer to nitrogen species in the heavy ends and B_0, B_1, and B_2 refer to nitrogen-containing compounds in the middle distillates.

The fractionation results presented in Tables IV and V, indicating a decrease in both the quantity and nitrogen content of the heavy ends in the product oils compared with their respective feeds, qualitatively support the above mechanism. The slight increase in the quantity of middle distillate fraction and the formation of light ends may be caused by the hydrogenolysis reactions. The nitrogen-containing polynuclear heterocyclics in the higher boiling fractions are being converted to those in the lower boiling fractions and the presence of nitrogen in the light ends of the product oils demonstrate the difficulty in denitrogenating the hydrodenitrogenation reaction network intermediate products.

Relative Activity

The hydrodenitrogenation data for MD and BD catalysts are plotted in Figure 7. These two catalysts have same metallic composition and are thought to be chemically similar. The nitrogen distribution data for two samples (one each obtained after 20 hr of operation) are plotted in Figure 8. The BD catalyst removed more nitrogen from all three fractions than the MD catalyst. The nitrogen contents in the light oils, middle distillates, and heavy ends for the BD catalyst were 0.039, 0.137, and 0.274 wt %, respectively; for the MD catalyst the nitrogen contents were 0.093, 0.172, and 0.398 wt %, respectively. The relatively high activity of the BD catalyst can be caused by the presence of a macropore system; hence, the pore diffusional resistances are affecting the hydrodenitrogenation activity of the MD catalyst. However, under the more severe operational conditions of 427°C (800°F), the MD catalyst appears to be better than the BD catalyst (Figure 9). Fractionation results plotted in Figure 10 indicate that the MD catalyst gives less nitrogen in all three fractions.

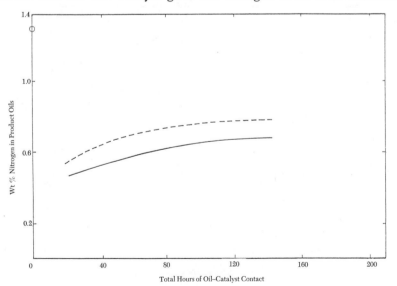

Figure 7. Comparison of catalysts for FMC oil hydrodenitrogenation. Catalyst: (– – –), MD; (———), BD. Temperature: 371°C (700°F); pressure: 1500 psig; space time: 2.99 hr (LVHST).

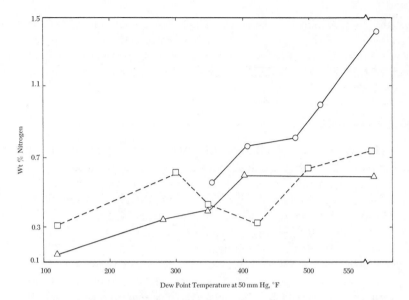

Figure 8. Nitrogen distribution in FMC product samples at 371°C (700°F). (○), Feed: FMC oil (7500 psig, and 2.99 hr (LVHST)). Catalyst: (△), BD and (□), MD.

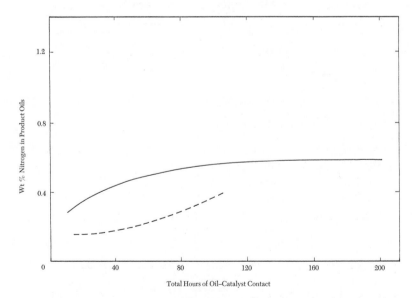

Figure 9. Comparison of MD (– – –) and BD (——) catalysts. Feed: FMC oil; temperature: 427°C (800°F); pressure: 1500 psig; space time: 2.99 hr (LVHST).

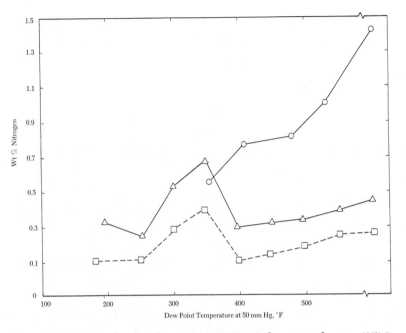

Figure 10.   Nitrogen distribution in FMC product samples at 427°C (800°F).   (○), Feed: FMC oil; (800°F), 1500 psig, and 2.99 hr (LVHST). Catalyst: (△), BD and (□), MD.

The total nitrogen in the product oils from runs using MD and BD catalysts were 0.205 and 0.483 wt %, respectively. For the BD catalyst the light oils, middle distillates, and the heavy ends contained 0.1, 0.186, and 0.194 wt % nitrogen, respectively; and for the MD catalyst the nitrogen contents were 0.034, 0.062 and 0.104 wt %, respectively.

Based on the results at 371°C (700°F), the BD catalyst would have been favored at 427°C (800°F), but the results indicate the superiority of the MD catalyst. One possible explanation is that the intrinsic activity of the MD is greater than that of the BD catalyst. At 427°C (800°F) severe hydrogenolysis would have occurred resulting in smaller nitrogen-containing species with apparently less pore diffusional limitations. An abrupt increase in the activity of the MD catalyst also was observed in the hydrotreatment of another heavy coal liquid, Synthoil, when the hydrotreating temperature was increased from 371°C (700°F) to 427°C (800°F) (13). In contrast to the experience with the heavy coal liquids, no significant difference in the activity of the catalysts was observed for raw anthracene oil. Raw anthracene oil has a narrower boiling range than the FMC oil and the heavy ends contain relatively less nitrogen than the middle distillates, whereas the distribution of nitrogen in FMC oil is quite different. These differences may be partly responsible for the observed variations in the performance of the catalysts while hydro-treating FMC oil and raw anthracene oil.

Deactivation

The nitrogen distribution in the product oil samples obtained during the operational intervals of 35–50 hr and 85–100 hr are plotted in Figure 11. Fractionation results indicated no significant difference in the amount of light oils at these two intervals; however, the heavy ends were initially 35% and were increased to 40 vol %. This suggests that the hydrogenolysis capability of the catalyst has decreased. The heavy ends also seem to contain more nitrogen in the product oils collected during 85–100 hr than in the earlier collected samples (Figure 11). This loss in activity may be caused by pore mouth plugging from carbonaceous deposits. The results plotted for the experimental run at 454°C (850°F) in Figure 2 reveal a high initial activity as well as a rapid decay rate. At this rather high temperature level significant coking occurs and it can effectively block the pore mouths. Most of the carbonaceous material was observed to deposit during the first 100 hr of operation (14) and the amount of carbonaceous material deposited on the catalyst surface at 454°C (850°F) was approximately twice that at the lower temperature

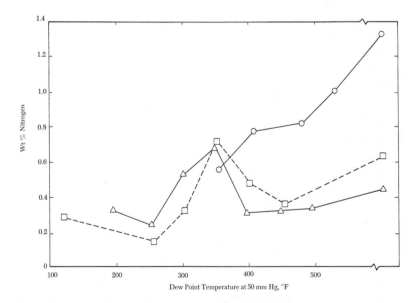

Figure 11.   Nitrogen distribution in product samples: BD catalyst. (O), FMC oil (427°C (800°F), 1500 psig, and 2.99 hr. (△), Product during 35–50 hr and (□), product during 85–100 hr.

levels. Hydrodenitrogenation activity decay also may be caused by the inhibitive adsorption of basic nitrogen-containing intermediates on the catalyst surface. This will result in decreased acidity of the catalyst surface and can be reflected in decreased activity for hydrogenolysis reactions and hence less nitrogen removal.

Conclusions

The following conclusions can be drawn from this study.

(A) Denitrogenation tends to be more difficult with an increase in the boiling range of the feed oil.

(B) Presulfiding seems to affect the catalyst activity in that a properly pretreated catalyst gives a high initial as well as stable activity.

(C) Under the reaction conditions of this study there seems to be no difference in the hydrodenitrogenation activity of MD and BD catalysts for hydrotreating raw anthracene oil.

(D) Apparently there are pore diffusional limitations in the hydrodenitrogenation of FMC oil, and the micro–macropore system in the BD catalyst seems to provide an advantage over the micropore system in the catalyst.

(E) Both MD and BD catalysts lose their hydrodenitrogenation activity while hydrotreating FMC oil. At higher temperatures this activity decreased at a significant rate. Deactivation can be the result of coking (pore mouth plugging) and inhibitive adsorption of basic nitrogen-containing compounds on the catalyst surface.

Literature Cited

1. Weisser, D.; Landa, S. "Sulfide Catalysts, Their Properties and Applications"; Pergamon: New York, 1973.
2. Satterfield, C. N., et al. *Ind. Eng. Chem., Process Des. Dev.* 1978, *17*(2), 141.
3. Shabatai, J.; Veluswamy, L.; Oblad, A. G. *Am. Chem. Soc., Div. Fuel Chem., Prepr.* 1978, *23*(1), 114.
4. Shih, S. S., et al. *Am. Chem. Soc., Div. Fuel Chem., Prepr.* 1978, *23*(1), 99.
5. Flinn, R. A.; Larson, O. A.; Beuther, H. *Hydrocarbon Process.* 1963, *42*(9), 129.
6. Mears, D. E. *Chem .Eng. Sci.* 1971, *26*, 1361.
7. Parakos, J. A.; Frayer, J. A.; Shah, Y. T. *Ind. Eng. Chem., Process Des. Dev.* 1975, *14*(3), 315.
8. Montagna, A. A., Shah, Y. T. *Ind. Eng. Chem., Process Des. Dev.* 1975, *14*(9), 479.
9. Satchell, D. P., Ph.D. Thesis, Oklahoma State University, Stillwater, OK, 1974.
10. Sivasubramanian, R., Ph.D. Thesis, Oklahoma State University, Stillwater, OK, 1977.

11. Ahmed, M. M.; Crynes, B. L. *Coal Process. Technol.* **1977**, *3*, 23.
12. Greenwood, G. J., Ph.D. Thesis, Oklahoma State University, Stillwater, OK, 1977.
13. Soni, D. S., M.S. Thesis, Oklahoma State University, Stillwater, OK, 1977.
14. Ahmed, M. M., M.S. Thesis, Oklahoma State University, Stillwater, OK, 1975.

RECEIVED June 28, 1978.

12

The Catalytic Effect of Active Metals upon Hydrodenitrogenation of Heavy Coal Liquids

CECILIA C. KANG[1]

Hydrocarbon Research, Inc. (a Subsidiary of Dynalectron Corporation), P.O. Box 6047, Lawrenceville, NJ 08648

A series of three catalysts were prepared for evaluating the effect of Co–Mo, Ni–Mo, and Ni–W upon hydrodenitrogenation of heavy coal liquids. The same support and method of preparation were used to prepare these catalysts. A heavy coal liquid derived from the solvent-refining coal process was used as the feedstock to evaluate these catalysts. Among these three metal pairs, Ni–Mo possesses the highest denitrogenation capability. The effects of temperature, pressure, and boiling range of feedstock were investigated. Any changes in operating variables which led to higher carbon deposition on the catalyst such as higher temperature, lower hydrogen pressure, and high resid-containing feedstock resulted in lower denitrogenation activity.

Several processes have been developed for coal liquefaction. Large-scale pilot plants have been in operation for the solvent-refining coal (SRC) process, and a pilot plant is being constructed for the H–Coal process, which is a direct catalytic process. Construction of demonstration plants is under consideration. The coal liquids produced from the current processes contain large amounts of residual fuels. They probably will be used initially as boiler fuels for stationary power plants. However, the nitrogen content of coal liquids is much higher than the petroleum residual fuels. The sulfur contents of coal liquids can vary considerably; they depend on the type of coal and the liquefaction process used. Current coal liquefaction processes are capable of produc-

[1]Present address: Catalysis Research Corporation, 450 Edsall Blvd., Palisades Park, NJ 07650

0-8412-0456-X/79/33-179-193$05.00/1
© 1979 American Chemical Society

ing fuels with sulfur contents sufficiently low to meet the legal limits, but the nitrogen contents of coal liquids may be too high to meet the NO_x emission limits.

The New Source Performance Standards set by the U.S. Environmental Protection Agency (EPA) for stationary sources specify a maximum nitrogen oxide emission from oil-fired furnaces at 230 ppm. With conventional low-nitrogen fuels, fixation of atmospheric nitrogen during the combustion process is the primary source of this emission. If organic nitrogen is present, roughly one-third of the organic nitrogen content of the fuel can react to form nitrogen oxides (NO_x) (the degree of conversion depends on firing methods). Five hundred parts per million of NO_x per percent of nitrogen in fuel oil is contributed when one-third of the nitrogen content of a fuel oil is converted to NO_x. This is over and above that from the fixation mechanism. Most of the coal liquids contain more than 1% nitrogen, hence further processing of coal liquids may be required to enable them to be used as boiler fuel for stationary power plants. The target set by EPA is maximum 0.3% nitrogen of coal-derived fuel oil.

At the beginning of 1976, EPA funded a laboratory program to develop an improved denitrogenation catalyst for heavy coal-derived fuel oil. The scope of this program consists of three tasks.

The objective of Task 1 is to compare the denitrogenation capability of three pairs of active metals (Co–Mo, Ni–Mo, and Ni–W) which are being used commercially for hydrotreating and hydrocracking of petroleum.

In general, the hydrodesulfurization capability, i.e., desulfurization with minimum ring saturation, of these three pairs of active metals are Co–Mo > Ni–Mo > Ni–W, whereas their hydrocracking capabilities are in the opposite order: Ni–W > Ni–Mo > Co–Mo. In view of the stable nature of the nitrogen compounds present in the coal liquids, catalysts possessing high hydrocracking capability might be desirable. Hence, a program is formulated to make a critical comparison of these three metal pairs on the same support (γ-alumina) and using the same method of preparation.

The objective of Task 2 is to improve the catalyst support; the purpose of Task 3 is to incorporate the best active metal pair onto the improved catalyst support. This presentation is devoted to the work performed for achieving the first objective.

Experimentation

Feedstock. SRC product and recycle solvent were obtained from the SRC pilot plant operated by Southern Services, Inc., in Wilsonville,

Table I.   Analysis of SRC Materials

	Material		
	SRC Solvent	SRC Product	SRC Blend
Gravity, °API	+1.8	−13.4	−6.8
Distillation (vol % at F°)			
ibp	378	700	403
10	463	945	516
30	539	—	605
50	605	—	730
70	658	—	1000
90	840	—	—
endpoint	935	—	—
Elemental analysis (wt %)			
carbon	88.35	86.39	87.76
hydrogen	7.65	5.91	6.80
sulfur	0.67	0.74	0.60
nitrogen	0.69	1.82	1.07
oxygen	—	4.29	3.30
ash (ASTM method)	—	0.16	0.09
total	97.63	99.31	99.62

Alabama. These were obtained from the processing of Illinois No. 6 coal from the Monterey Mine. The analysis of SRC materials is presented in Table I.

A 60/40 blend of recycle solvent/SRC was prepared to yield a feedstock pumpable at 150°F. A detailed characterization of the feedstock is given in Table II.

Table II.   Inspection of SRC Feedstock

	Total Liquid	ibp–975°F	975°F+
Wt %	100	65.3	34.7
Gravity (°API)	−6.8	—	—
Carbon (wt %)	87.76	88.09	87.56
Hydrogen (wt %)	6.80	7.46	5.61
Nitrogen (wt %)	1.07	0.77	1.84
Sulfur (wt %)	0.60	0.32	1.11
Oxygen (wt %)	3.09	2.32	3.50
Ash (wt %)	—	—	0.03
Total (wt %)	99.32	98.96	99.65
Benzene insoluble (wt %)			24.49
Benzene soluble (wt %)			75.51
Viscosity (cP at 200°F)	85		
Pour point (°F)	70		

Catalyst Evaluation. Experiments were conducted in a stainless steel reactor containing 150 mL of catalyst. The catalyst bed measured 15 in. in length. A 3-point thermocouple was placed in the middle of the catalyst bed to monitor the temperature at the inlet, center, and outlet of the reactor. A lead bath was used to heat the reactor, attempting to achieve an isothermal operation, which frequently was attained. However, temperature rises were observed when good performance was achieved.

The catalyst was presulfided at operating temperature for 1 hr using more than twice the amount of H_2S needed to convert the metals to sulfides. Reaction conditions were 780°–850°F, 2000–2800 psig, 0.5 VHSV, and excess hydrogen of 4000 scf/bbl. A standard shutdown procedure was used. Anthracene oil was flushed through the reactor at 650°F for 1 hr.

The tests were conducted for a 3-day period. Gas products were taken and analyzed by mass spectrometry (MS). Daily liquid samples were taken and their nitrogen contents were determined. The liquid product taken at the end of the test period was characterized by elemental analysis, distillation, and benzene extraction. A representative sample of the spent catalyst was obtained by rifling, then washing with benzene. Its carbon content was determined.

Catalysts. A batch of γ-alumina extrudates was prepared. Active metals were incorporated by impregnating the extrudates with corresponding solutions of the metal salts. The characteristics of the catalysts are presented in Table III.

Table III.   Characteristics of Catalysts

Property	Units	SN-4412A	SN-4412B	SN-4412C
NiO	%	3	—	3
CoO	%	—	3	—
MoO$_3$	%	15	15	—
WO$_3$	%	—	—	15
Diameter	in.	0.065	0.065	0.065
PV (H$_2$O)	mL/g	0.84	0.85	0.88
CBD	g/mL	0.54	0.54	0.51
Surface area	m^2/g	204	195	237

Results and Discussion

Comparison of Active Metal Pairs. The catalysts were evaluated at a temperature of 780°–850°F at a 2000°–2800°F pressure with a space velocity of 0.5 vol/vol/hr. The range of product distribution from this evaluation is summarized in Table IV. The quality of liquid product as defined by gravity, and by residuum (975°F +), hydrogen, nitrogen, sulfur, and oxygen contents, and by carbon on spent catalysts is presented

Table IV.   Product Distribution[a]

$CO + CO_2$	0.1–0.4
$H_2S + NH_3 + H_2O$	3.8–6.7
C_1–C_3	1.6–6.4
C_4–C_6	1–4
ibp–975°F	71–84
975°F+	10–20

[a] Wt % output basis.

in Table V. The data indicate that Ni–Mo has higher denitrogenation capability than Co–Mo and Ni–W.

Temperature Effect. The effect of temperature upon the performances of the Co–Mo, Ni–Mo, and Ni–W catalysts and the carbon content of the spent catalysts are shown in Tables V and VI and Figures 1, 2, and 3. Carbon deposition on the catalysts is excessive, much higher than that encountered during hydroprocessing of coal under similar operating conditions (1). A 10–15% decrease in the absolute carbon contents of the spent catalysts was observed by a decrease of the operating temperature from 850° to 810°F, accompanied by significant improvements in removal of heteroatoms and 975°F+ conversion for Co–Mo and Ni–Mo catalysts. The temperature effect upon the performance of the Ni–W catalyst is mixed. This type of temperature effect does not occur in hydroprocessing of coal.

As shown in Table VI, for the Ni–Mo catalyst, further decrease of the operating temperature from 810° to 780°F gave poor liquid product quality, coupled with no further decrease of carbon deposition on catalyst. Hydroprocessing of the SRC solvent under the same conditions as those used for the processing of the 60/40 blend of solvent/SRC gave much deeper denitrogenation than that achieved for the blend containing the heavy SRC product. Carbon deposition on the Ni–Mo catalyst resulting from the hydroprocessing of the solvent is very low. During the 3-day testing period, the denitrogenation activity of the catalyst remained at the 87% level for the solvent but decreased from 77 to 66% for the blend. The derived data indicate unequivocally that the major factor contributing to the initial deactivation of the catalyst is carbon deposition on the catalyst, which is contributed heavily by the heavy coal liquids.

Pressure Effect. The pressure effect upon these three catalysts is shown in Figure 4. A decrease in operating pressure from 2800 to 2000 psig gave less change in liquid product quality than the decrease of operating temperature from 850° to 810°F. Overall, the decrease in pressure caused less hydrogen consumption, higher 975°F+, and lower hydrogen content of the liquid products, accompanied by higher carbon deposition on the Ni–Mo and Ni–W catalysts, but no change in carbon deposition on the Co–Mo catalyst.

Table V.   Denitrogenation of 60/40 Blend

Temperature (°F)		850		
Pressure (psig)		2800		
Space velocity (vol/vol/hr)		0.5		

| | | Catalyst | | |
		Co–Mo	Ni–Mo	Ni–W
H₂ Consumption (scf/bbl)		990	3010	2480
Liquid product	Feed			
gravity (°API)	−6.8	2.7	11.6	8.7
975°F+	34.7	17.8	19.2	11.5
975°F+, benzene insoluble	8.5	6.0	0.7	0.7
hydrogen (wt %)	6.8	7.4	9.0	8.6
nitrogen (wt %)	1.07	0.65	0.39	0.60
sulfur (wt %)	0.60	0.15	0.21	0.03
oxygen (wt %)	3.09	1.1	0.61	0.77
Carbon on spent catalyst		39.1	31.7	27.1

Table VI.   Denitrogenation of 60/40 Blend of Recycle

Temperature (°F)		850	810
H₂ consumption (scf/bbl)		3010	2890
Feedstock	Blend		
Liquid product			
gravity (°API)	−6.8	11.6	13.5
975°F+ (wt %)	34.7	19.2	10.6
975°F+, Benzene in-			
soluble (wt %)	8.5	0.7	0.1
hydrogen (wt %)	6.8	9.0	9.5
nitrogen (wt %)	1.07	0.39	0.37
sulfur (wt %)	0.60	0.21	0.04
oxygen (wt %)	3.09	0.61	0.34
Carbon on spent catalyst		31.7	21.5

[a] Ni–Mo catalyst, 2800 psig, 0.5 VHSV.

of Recycle Solvent/SRC (Monteray Coal)

	810 2800 0.5			810 2000 0.5		
	Catalyst					
	Co–Mo	Ni–Mo	Ni–W	Co–Mo	Ni–Mo	Ni–W
	2640	2890	2340	2220	2070	1760
	10.2	13.5	7.7	7.6	7.5	6.5
	11.6	10.6	15.8	15.5	14.8	16.5
	0.1	0.1	0.2	0.2	0.4	0.3
	9.1	9.5	8.8	8.5	8.4	7.9
	0.54	0.37	0.70	0.48	0.41	0.63
	0.05	0.04	< 0.03	< 0.03	< 0.03	< 0.03
	0.35	0.34	0.37	0.18	0.33	0.19
	24.2	21.5	15.8	23.6	30.4	21.1

Solvent/SRC and Recycle Solvent (Monterey Coal)[a]

	780 1900	*SRC* *Solvent*	*780* 3220
	5.8	1.8	23.7
	20.1	0	0
	0.4	0	0
	8.8	7.7	11.4
	0.55	0.69	0.09
	0.08	0.67	0.05
	0.30	0.27	—
	23.1		10.7

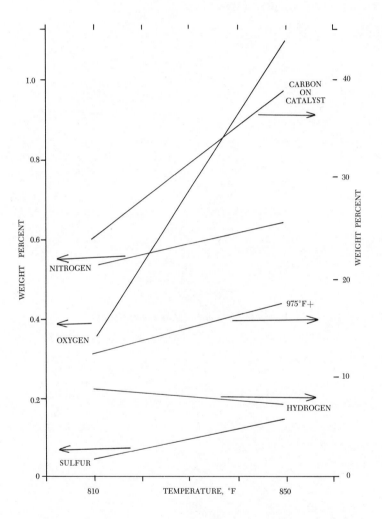

*Figure 1.   Temperature effect upon Co–Mo catalyst at 2800 psig
and 0.5 VHSV*

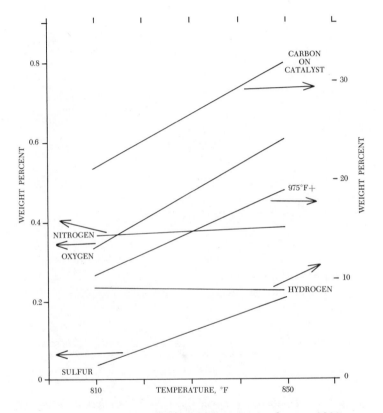

Figure 2. Temperature effect upon Ni–Mo catalyst at 2800 psig and 0.5 VHSV

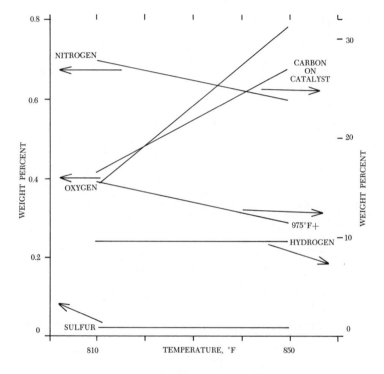

Figure 3. Temperature effect upon Ni–W catalyst at 2800 psig and 0.5 VHSV

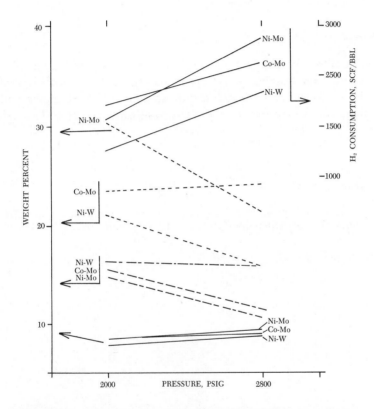

Figure 4. Pressure effect at 810°F and 0.5 VHSV: (– – –), carbon on catalyst; (— – — –), 975°F+ in liquid product; (——–), hydrogen in liquid product.

Conclusions

For denitrogenation of coal-derived fuel, Ni–Mo is more effective than Co–Mo, and Co–Mo is better than Ni–W. The initial catalyst deactivation is caused mainly by carbon deposition. Higher temperature operation resulting in higher carbon deposition on the catalyst gives poorer performances in terms of denitrogenation, desulfurization, deoxygenation, and 975°F+ conversion.

Acknowledgments

The author wishes to acknowledge the U.S. Environmental Protection Agency for funding of this program, R. H. Ebel and his staff at American Cyanamid for catalyst preparation, and J. Gendler and P. Maruhnic for their assistance in conducting experiments and compiling data.

Literature Cited

1. Kang, C. C.; Johanson, E. S. "Deactivation and Attrition of Co–Mo Catalyst during H-Coal Operation," in "Liquid Fuels from Coal"; Academic Press, 1977; pp. 89–101.

RECEIVED June 28, 1978.

INDEX

Jacket design by Carol Conway.
Editing and production by Candace A. Deren.

*The book was composed by Service Composition Co., Baltimore, MD,
printed and bound by The Maple Press Co., York, PA.*